SECURITY AND DEFENCE IN SOUTH-WEST ENGLAND BEFORE 1800

Edited by Robert Higham

Produced in conjunction with

The Centre for South-Western Historical Studies

University of Exeter

EXETER STUDIES IN HISTORY No. 19

First published by the University of Exeter 1987.

EXETER STUDIES IN HISTORY

ISBN 0 85989 209 3
ISSN 0260 8628

Printed and bound in Great Britain by
Short Run Press Ltd., Exeter

Contents

List of figures

Acknowledgements

The contents of this volume were given originally as papers at the first Symposium of the Centre for South-Western Historical Studies, held in Exeter in November 1986, before an audience of about one hundred people. The opening event of the Symposium was the eighteenth Harte Lecture on Local History, given by Professor Joyce Youings, Professor of English Social History at Exeter University. We gratefully acknowledge the assistance of the organisers of the Symposium for their help with the production of this volume, which we hope will be only the first of a number of 'South-Western' issues in the *Exeter Studies in History* series.

Preface

BY ROBERT HIGHAM

No society has ever been truly free from threat. From the earliest times until the present day, social groups both large and small have been subject to problems of security posed by outside enemies as well as by internal conflicts and tensions. The responses to these problems are seen in various ways, for example, in the raising and deployment of armed forces, in the securing of frontiers and in the building of fortifications. One response might stem from the top of a heavily governed society, in the form of a modern standing army, for example. Another might stem from a more locally-based authority in a more decentralised society, in the form of a medieval castle, for example. Some problems of security and defence are posed by major and obvious outside threats, commonly from a neighbouring country. Others are posed by weaknesses inherent in the society concerned, either specific—as in a dynastic civil war—or more general—as during much of the medieval period when political authority was more fragmented than it was later to be.

Although the issues of security and defence are commonly examined at national level, there is a strong case for arguing that a regional framework is more appropriate, especially before the contemporary period. The practicalities of security and defence are heavily influenced by local geography, for example: by communications, by proximity to or distance from other areas, by the nature of the terrain, by land-locked or maritime situations and so on. The South West of England provides a particularly good example of the importance of such local factors. Its peripheral character and often difficult terrain have meant slow and often partial influence from societies further east. But its peninsular shape and long coastline have made it accessible to sea-borne threat.

The contrasting themes of internal remoteness and external vulnerability, which have been so crucial to the region's security in the past, are explored in a number of ways in the essays in this volume. For the Roman period, Valerie Maxfield considers the problem of security and defence from the point of view of an invading force whose aim was to conquer the area and control it. The variety of responses—mainly in the form of fortifications—which different societies in the Middle Ages made to a problem of security which had both internal and external dimensions, is the theme of Robert Higham's contribution. Joyce Youings analyses the particular problems of raising and organising the local militia in the Tudor period. The two seventeenth-century chapters—by Ivan Roots and Anne Duffin—provide a close-up view of Cornwall early in the century and a wider view of the region during the Commonwealth and Protectorate. Finally David Starkey, for the eighteenth century, explores the role the region played in the North Atlantic fishing industry, whose contribution to the manning of the British Navy marked the interplay of trade and security at local and national levels.

Throughout, the assumption of the authors is clear that security and defence, while constituting what was literally a life-and-death issue for the populations of the south-western peninsula, formed part of wider patterns of conflict and dependence which linked the region to the outside world. Prior to 1800 at least, the national and international themes of security and defence were played out most cogently at the regional level.

The Army and the Land in the Roman South West

BY VALERIE A. MAXFIELD

The army with which this paper is concerned was an army of conquest. In this respect it differs significantly from the other military bodies dealt with in this volume, all of which were national institutions concerned with the defence and the maintenance of security against hostile foreigners. The Roman army constituted such a foreigner, invading Britain with a view to absorption of the island into the Roman world. Some time during the middle 40s AD the army came to the south-west of Britain and remained there just as long as was necessary to complete the conquest. In the 70s it moved on, westwards into Wales, northwards into Scotland: the whole south-western area, in common with the rest of southern and eastern Britain, was demilitarised and administration of the newly-won area was handed over to local self governing bodies, *civitates*, city states. The period with which this paper is concerned is thus a relatively brief one, little over 30 years in total; its theme is the process by which the Roman invader established and maintained security and the nature of the military presence which established that security.

The sources of evidence are three-fold, literary, archaeological and epigraphic, tantalising snippets of information thrown out by Roman historians, combined with a study of the evidence on the ground (the settlements of the natives and the military bases of the invaders) and the occasional tombstone commemorating a member of the invading army. There is none of the contemporary documentation available for later periods which allows of a detailed analysis of the composition of the armies and permits the construction of a relatively detailed account of military affairs. Necessarily, therefore, much of what follows is in the nature of conjecture and surmise, based, all too often, on the evidence of analogy. In the absence of a coherent historical narrative it is necessary to use mute archaeological evidence in order to try to understand the methods and motives of the conquerors, to elucidate the attitudes and reactions of the natives.

As already indicated, the historical sources relevant to our theme are sparse and, at best, second-hand. Those chapters of the Annals of Tacitus which will have dealt with the Claudian invasion of Britain are missing; all that has come down to us is the History of Cassius Dio, a Greek historian writing nearly two centuries after the events he records.[1] His concern is with the initial landing in the South East, the immediately ensuing encounters, the arrival of the emperor Claudius and the triumphal entry into Camulodunum (Colchester). Subsequent events are dismissed in the laconic statement that Claudius ordered his general, the senator Aulus Plautius, to subdue 'the rest'; the rest of what we are not told, presumably the rest of the island of Britain, though many subsequent commentators have taken the term to mean no more than the rest of lowland Britain. The Tacitean narrative takes up some four years later, but its concern is almost exclusively with events outside the South West, though it does touch, at one point, on matters relevant to its periphery.[2] The most useful, albeit infuriatingly imprecise, allusion to the South West comes in the pages of the anecdotal imperial biographer Suetonius Tranquillus, whose life of the emperor Vespasian includes a summary account of the future emperor's exploits in Britain in the 40s, where he was commanding legion *II Augusta.*[3] Thus we are told that while in Britain Vespasian conquered two warlike tribes, fought thirty battles, took twenty *oppida* and the Isle of Wight. The reference to Vectis, the Isle of Wight, provides the one geographical peg onto which these exploits can be hooked, though it is assumption and not established fact, that the two tribes, the thirty battles and the twenty *oppida* are all to be located in the southern and western part of Britain. The Roman army was a highly mobile creature and during the time that Vespasian remained in Britain the troops under his command could have operated over a very considerable stretch of terrain. *II Augusta* was one of the four legions which formed the core of the invasion army, Vespasian bringing it over from its previous base at Strasbourg in the upper Rhineland and remaining with it as its commander for a period of about three years.[4] The association of legion II with the South West is reinforced by an allusion in the Geography of Ptolemy. Ptolemy, writing in the middle years of the second century AD, locates the legion at Isca, which he includes among the *poleis* of the Dumnonii; that is to say at Exeter, Isca Dumnoniorum. It is certain that he is wrong in the sense that at the time at which he was writing, legion *II Augusta* was stationed at Caerleon in South Wales, that is at Isca Silurum; but an element of truth may be detected behind this error, a confusion rather than a falsehood. Ptolemy drew on a variety of sources for his information, and it is clear that for southern Britain these included first century material. Hence a confusion between a first century attribution of legion II to Isca Dumnoniorum with the knowledge that in the second century its base was Isca Silurum.[5] There is no epigraphic evidence to attest the presence of the legion at Exeter, but a decorated antefix from the demolition deposits in the legionary bathhouse, which is paralleled exactly by an example from Caerleon, provides convincing enough confirmation.[6] Another of the four invasion legions, legion XX, is also attested in the South West, but doubt surrounds the identification of its base. In the winter of 49/50 it moved westwards, out of its previous fortress at Colchester into a position from which it was in touch with the Silures of south Wales. A fragmentary and now lost inscription from Gloucester, the tombstone of a serving soldier of the legion, suggests that it (or part of it) may have been based here at one time.[7] The legionary fortress which underlies the medieval and modern city is known not to have been founded until the late 60s, but another military site lies to the north at Kingsholm, and material from here suggests occupation ranging from the Claudian period up to the latter part of the 60s. Despite a recent thorough re-examination of the evidence from the site, nothing positive can be said about its size and hence its capacity (a complete legion requires something in the region of 20 hectares).[8] There appear to have been two separate building phases at Kingsholm, distinguishable from one another by the use of different structural techniques, post-in-trench and sill-beam construction. Post-in-trench buildings extend over an area of something between 2.1 and 8.05 hectares, while buildings on the Kingsholm alignment are estimated to cover a total of at least 17.9 and up to 22.4ha. There is no way of knowing with which phase the legionary tombstone is to be associated. Another site which has been associated with legion XX is Usk, a 19.4-hectare base founded at some time during the 50s. A small bronze roundel decorated with the symbol of a boar, the totem of legion XX, was retrieved from the excavations here.[9] A legion based at either of these sites will have been well positioned to operate among the tribes of Wales and both are probably to be connected with the campaigning in these western areas, the liberal granary provision at Usk suggesting that this site perhaps served primarily as a campaign supply base. Whether or not legion XX operated at all in the South West is unknown. All that the archaeological evidence can tell us is that it was based on the periphery of the area: the reason for this, according to Tacitus, lay with campaigns beyond.

In addition to the legions, the invasion force was comprised of auxiliary troops, non-citizen soldiers enrolled in units nominally 500 strong, composed of cavalry, infantry, or a mixture of the two. The number of auxiliaries who came over in 43 is not known with any precision. The common assumption that there will have been a comparable number of legionaries and auxiliaries, that is about 20,000 of each (the equivalent of forty auxiliary regiments) is quite feasible but incapable of proof. It is known that by AD 122 there were fifty units in Britain some of which were, by this time, 1000-strong.[10] Allowing for some departures and arrivals during the intervening years (departures for the East in 66, for the civil wars in 69, for the Danube in 86 and again in 101: reinforcements after the Boudiccan rebellion in AD 60-61, for the Flavian campaigns launched in AD 71 and for the northern wars in the early second century), a figure of about forty units in the Claudian army appears quite moderate, though it has recently been argued that a large proportion of the auxiliary units subsequently attested in Britain came over not in AD 43 but with Cerealis in 71.[11] The basis for this belief is the fact that comparatively few units are specifically attested in Britain during the first thirty or so years of the occupation. While it certainly is true that many units cannot be proved to have come over to Britain with the invading army, nor can many of them be proved to have been elsewhere at this period. The problem arises from the nature of the evidence for the movements of troops, the most important sources being military diplomas, tombstones, building inscriptions and stamped tiles. We have no diplomas for the British army earlier than the Trajanic period.[12] Tombstones are comparatively rare in lowland Britain, that is that area of Britain in which the army was based prior to the Flavian advances, and at that period auxiliary soldiers are not known to have been involved in building operations which were normally carried out by legionary troops. Moreover the inscriptions erected at turf and timber forts will, no doubt, often have been of timber not stone and will therefore not have survived. It is hardly surprising therefore that few of the units which came over in 43 can be individually identified and located. Four units can however be pinpointed in the South West, all of them in the Avon/Gloucester region (that is in the area of the Jurassic ridge which provided excellent, easily worked stone for funerary purposes). The evidence in each case is that of a tombstone. At Cirencester two cavalry units, the *ala Gallorum Indiana* and the *ala Thracum* are attested, the latter unit almost certainly to be equated with a unit of the same name attested at Colchester.[13] It is highly probable that it moved west with legion XX in 49/50. The order in which these two units occupied the Cirencester fort is a matter for speculation but of no particular import in the present context.[14] A trooper from the *cohors VI Thracum equitata* was buried at Gloucester, and may well have shared accommodation at the Kingsholm fort with the legionary soldier attested from the same cemetery, though equally they may relate to the two separate phases of the site.[15] Finally there is the tombstone of a cavalryman from the *ala Vettonum* at Bath. The interpretation of this stone has recently been debated, the point at issue being whether the trooper in question, one Lucius Vitellius Tancinus, died at Bath when his unit was stationed there, or whether he came to Bath to take the waters at a time when the unit itself was stationed elsewhere, specifically at Brecon Gaer in south Wales where it is attested in the late first century.[16] The crux of the problem lies with the dating of the inscription, whether it is Claudio-Neronian, in which case a base at Bath is probable, or Flavian, in which case it is not. The case for a Claudio-Neronian date rests on the interpretation of the soldier's names. The form in which the nomenclature is given on the tombstone, *L. Vitellius Mantai f. Tancinus* would appear to indicate a man with the *tria nomina* of a Roman citizen, who is the son of a non-citizen, Mantaius. The obvious implication of this is that Tancinus won his citizenship in his own right rather than inheriting it from his father. There are two possible contexts. Since he had served for twenty-six years, Tancinus could have profitted from the grant of citizenship which, from the time of Claudius onwards, was awarded to all auxiliary soldiers on completion of twenty-five or more years service. Alternatively he could have won it on the occasion when the unit in which he was serving, the *ala Hispanorum Vettonum civium Romanorum*, received a block grant of citizenship as a reward for distinction in battle. This practice of awarding the battle-honour *c.R.* was very common from the time of Vespasian, but may well have been introduced by Claudius who is known to have been generous in his grants of Roman citizenship. An encounter during the invasion of Britain would provide a suitable occasion for such a grant, and it is tempting to associate it with the award of military decorations to an equestrian officer, one Stlaccius Coranus, whose career inscription records that he was honoured at some time during a military career which ended as prefect of the *ala Vettonum.*[17] Whatever its precise context, the date of Tancinus's acquisition of Roman citizenship is indicated by the *praenomen* and *nomen* which he assumed, Lucius Vitellius, highlighting an award made during the censorship of the emperor Claudius and Lucius Vitellius, that is AD 47-48. This being so the tombstone in question dates somewhere between 47 and 72 with a bias towards the earlier rather than the later end of this range. Such an early date is, however, doubted by Holder, who believes that the whole practice of making block grants of citizenship to auxiliary units, *ob virtutem*, dates no earlier than the Flavian period, and who hence adduces an alternative interpretation of Tancinus's names. He argues that although Tancinus's father appears to have only the single name of the non-citizen, he was in fact a citizen, Lucius Vitellius Mantaius, but is referred to by his *cognomen*, Mantaius, instead of by the more normal form of filiation using the *praenomen*, Lucius. Hence it is the father, Mantaius, and not the son, Tancinus, who acquired citizenship during the censorship of Lucius Vitellius: the son need therefore have served and died no earlier than the Flavian period when the *ala Vettonum* is known to have been stationed in Wales. All association between the unit itself and an assumed base at Bath therefore disappears. The case cannot be proved one way or the other, so in the absence of evidence to the contrary it would seem more satisfactory to accept the more conventional interpretation of the text, particularly in view of the fact that the Holder hypothesis presupposes the enlistment of a Roman citizen into a non-citizen unit — an unlikely occurence at the date in question. [18]

All these soldiers known or believed to have been stationed in the South West in the early first century were, of course, alien to the area, having been recruited into the army before it came to Britain. We can identify (using modern place-names) one Bulgarian, one Swiss, one Portuguese and one native of the Low Countries. However, it was not long before Britons started to join the ranks of the Roman forces. A diploma of AD 80 refers to soldiers in a unit of Britons who had served for at least twenty-five years and must therefore have joined up in 55 at the latest, while British tribesmen certainly took part on the Roman side in the battle against the Caledonian leader, Calgacus. A Dumnonian (native of Devon or Cornwall) served in the *classis Germanica*, the Roman fleet on the Rhine, perhaps as early as the 80s, while a citizen of Winchester and another of Cirencester took part in Trajan's Dacian wars in the early years of the second century AD.[19]

The participation of Britons on the Roman side fighting against other Britons in the battle of Mons Graupius is particularly significant, for it points to a factor of outstanding importance for our understanding of the nature of the Roman conquest, and hence of our interpretation of the archaeological evidence relating to it. That factor is the tribal nature of British society in the late pre-Roman Iron Age. The conquest of Britain was not the conquest of a single united enemy, it was the conquest of an ill-assorted collection of tribal units, unable to sustain for long a united opposition against a common external enemy. Indeed, not all of the tribes of Britain were hostile to Rome; some were clearly philo-Roman, for in a society with endemic inter-tribal strife some will have seen in Rome a support, an external ally against an internal aggressor. It was just this kind of inter-tribal strife resulting in the expulsion of the Atrebatan king Verica (or Berikos as

Dio calls him) which served as a pretext for the whole invasion of Britain. This internecine strife was exploited by Rome for her own ends, and within a few years of the initial invasion alliances had been cemented with rulers in East Anglia (Prasutagus of the Iceni), northern England (Cartimandua of the Brigantes), central southern England (Cogidubnus of the Atrebates) and, if Dio's 'Bodunni' are correctly identified as the Dobunni, an area on the eastern bank of the lower Severn. Since the previous king of the Atrebates, Verica, had been expelled immediately prior to the invasion, it is not improbable that Cogidubnus was a puppet ruler installed by Rome. We know that (in common with so many of Rome's allies) Cogidubnus was rewarded with Roman citizenship: an inscription from Chichester terms him *rex magnus*, great king, and names him as Tiberius Claudius Cogidubnus, indicating citizenship awarded by one of the Claudii, very probably the emperor Claudius himself.[20]

While the friends of Rome are identifiable by name the enemies are not and it is to archaeology that we must turn in an attempt to identify the 'two warlike tribes' to which Suetonius alluded. The region of the Durotriges provides the most graphic evidence. The war cemetery at Maiden Castle, the mass burial at Spettisbury, the artillery barrage and fort construction at Hod Hill, all point to encounters with Roman troops. The supposed artillery barrage at this last site may however be questioned. The contours of the hill make the aiming of catapults at a single specific building (interpreted as the chieftain's house) as suggested by Richmond highly improbable, and it is not unlikely that the artillery bolts derive from the activities of the Roman troops subsequently based at the fort; the erstwhile hill-fort site will have made an excellent practice ground. Isolated finds of bolts shot from Roman catapaults at Pilsden Pen and at an earthwork near Badbury Rings are rather less convincing indicators of assault, while judgement on the context of the South Cadbury material (a burnt gate, unburied bodies and Roman military-type buildings within the native hill-fort) must be reserved until such time as the evidence for the proposed chronology is published.[21] The problem of dating such military encounters from archaeological evidence is a difficult one; if distinctively Roman material is present (as in the case of the Spettisbury and Cadbury assemblages) we can at least postulate a context within the invasion period; when no such material is present such an interpretation necessarily becomes more speculative. The Worlebury massacre is a good case in point. Excavation here in the last century produced parts of eighteen skeletons, nine with evidence of weapon damage. One of the bodies had been decapitated, others exhibited what appeared to be sword cuts on the skulls. All had been apparently unceremoniously dumped in storage pits.[22] This material is often quoted as evidence of an encounter between Roman and Briton, but equally it is used as evidence of pre-Roman inter-tribal warfare, an episode in the strife between the native Dobunni and Belgic invaders.[23] While the assemblage clearly comes from the latest Iron Age occupation at the site, it cannot be dated more precisely than that; hence the exact historical context is a matter of opinion not of firm evidence. The temptation to use the advent of Rome as a convenient peg on which to hang various observations relating to the final Iron Age phases of defended sites is a strong one. In his early interim statements on the final pre-Roman refurbishing of the defences of Danebury, Cunliffe tentatively suggested it might be a response to the threat of Rome. His suggestion was carefully hedged around with reservations, reservations which were dropped by subsequent commentators who turned a working hypothesis into a statement of fact. The final report on the Danebury excavations discards the hypothesis entirely.[24] The final pre-Roman phase is established as being well pre-Roman; the site was deserted when Roman troops came by. Indeed Cunliffe has argued, with particular reference to the area of central southern England, that in the later Iron Age many hillforts were abandoned, with hilltop occupation being concentrated in a few more widely scattered sites.[25] How far this picture can be extended down into the South West is not clear, and will become apparent only with further detailed excavation, but Todd notes in an interim statement on his excavations at Hembury in Devon, the latest hillfort investigation in the South West, that the site was abandoned within the first decade BC: the Roman fort constructed there in the mid-first century AD was constructed within a convenient site empty of occupation.[26]

While the Durotriges may reasonably be taken to have been one of the tribes hostile to Rome, the attitude of the Dumnonii is impossible to determine. Literary evidence is totally lacking and the archaeological evidence is capable of more than one interpretation. As already noted, the chronological relationship of Roman to native fortification at Hembury is such as to rule out any notion of hostile encounter here. The refortification of the gateway of the hillfort at St Mawgan-in-Pyder, dated to c.50-70, has been associated with the westward advance of the Roman army, but the very fact that the site remained in occupation, its defences apparently intact, well into the second century AD suggests that it constituted no threat to the invaders.[27]

The extent to which it is ever going to be possible to determine a 'tribal attitude' towards Rome depends, of course, on whether or not any such single attitude ever existed, whether there was a united response, hostile or friendly, to the invader. A unified response presupposes the presence of either a charismatic war-leader capable of temporarily welding together disparate groups for the purposes of defence or of the pre-existence of a power structure capable of coordinating such a response. It is at least questionable whether such a structure existed in the more westerly part of the south-west peninsula at that time. The political advances made by the more easterly British tribes in the century or so before the Roman conquest, leading to the concentration of power, the emergence of 'kings', the use of coinage sometimes naming these leaders, the abandonment of many fortified hilltop sites and the drift towards powerfully defended but lower and more centralised sites, are phenomena which had barely impinged on the Dumnonii. The existence of something of a 'tribal identity' is suggested by the very existence of a tribal name recognizable by Rome and incorporated into the title of the *civitas*, Isca Dumnoniorum, Isca of the Dumnonii, but to what extent and in what ways this was reflected in the pre-Roman political and military structure in the region is not at all apparent. No coinage was minted, nor is there any indication of any major central focus or foci within the area. There exists a multiplicity of small defended enclosures which might suggest a fragmentation of power if one but knew how many of these sites were in contemporary occupation, when and by whom. But we do not.

Another approach to the problem is via the evidence of the Roman military structures which appear in the area. At its most simple level it has been argued that where Roman forts are present the natives were hostile; where no forts appear the natives were friendly. This argument has been applied to the south-western peninsula which, until comparatively recently appeared to be devoid of all signs of Roman military occupation. Now that many sites are known the pendulum has tended to swing in the other direction. An absence of forts does not necessarily mean that an area put up no resistance: if that were so, archaeologically speaking Caesar never conquered Gaul ! But nor does a multiplicity of forts necessarily imply fierce resistance and the necessity to 'hold down' the natives, a function commonly attributed to a Roman fort. Such a simplistic approach to the evidence ignores a number of important factors which fundamentally affect our understanding of the situation.

Firstly, there is the very basic point that an archaeological distribution map (such as Fig. 1) provides no more than an indication of the state of knowledge at the date when it was drawn up. Fig. 1 includes all military sites of known or assumed pre-Flavian origin in the South West: there are twenty-six sites classed as 'certain', eight that are 'probable', giving thirty-four in total. The Ordnance Survey map of Roman Britain published in 1956 knows of only six of these, but includes, in addition, one questionable case (Nettleton Shrub), the excavator of which revised his interpretation of

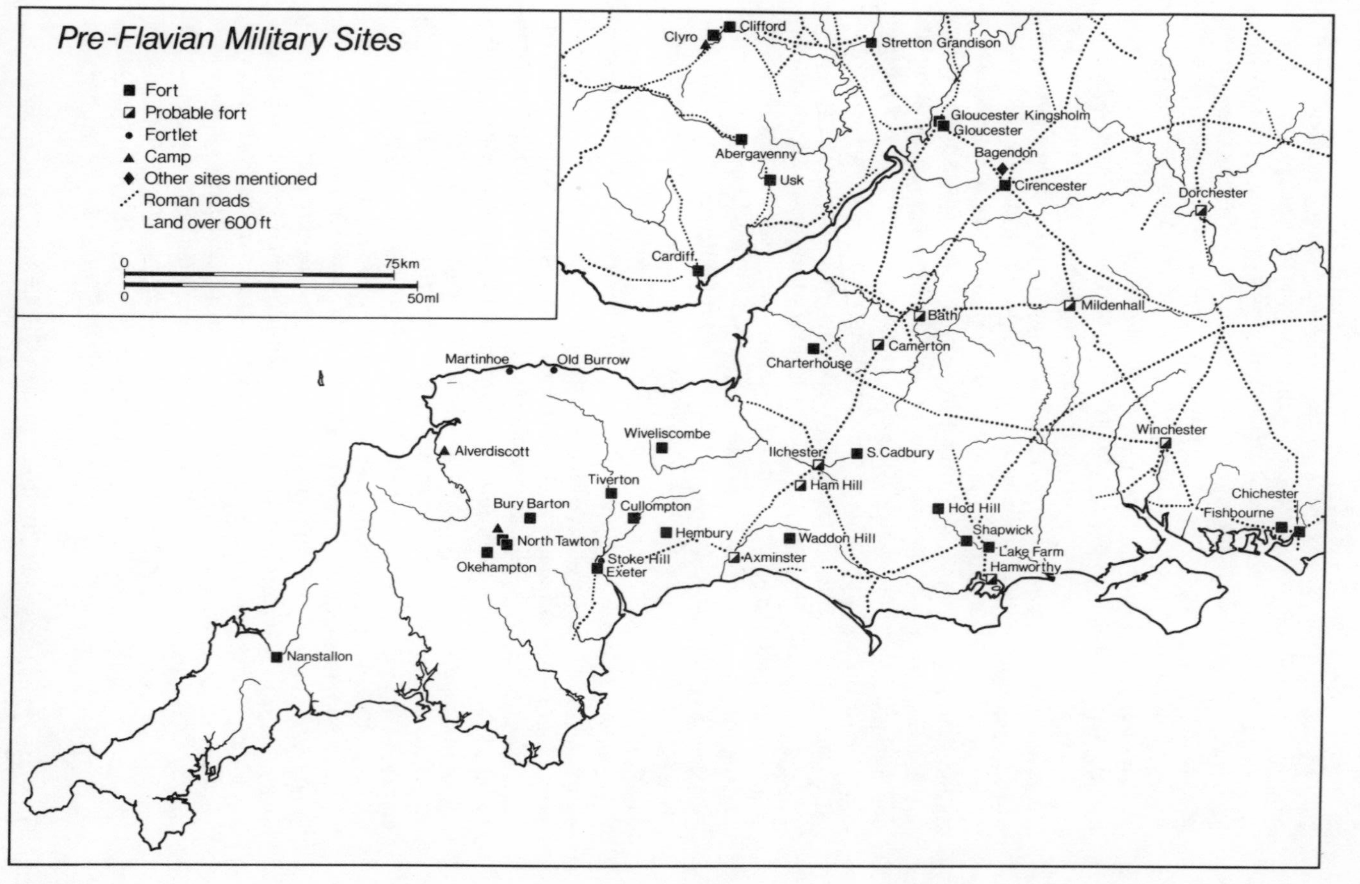
Pre-Flavian Military Sites
Fort
Probable fort
Fortlet
Camp
Other sites mentioned
Roman roads
Land over 600 ft
0
75km
0
50ml
Clyro
Clifford
Stretton Grandison
Gloucester Kingsholm
Gloucester
Abergavenny
Bagendon
Usk
Cirencester
Dorchester
Cardiff
Bath
Mildenhall
Camerton
Charterhouse
Martinhoe
Old Burrow
Wiveliscombe
Alverdiscott
Ilchester
S.Cadbury
Winchester
Tiverton
Ham Hill
Chichester
Bury Barton
Cullompton
Hod Hill
Fishbourne
Hembury
Shapwick
North Tawton
Waddon Hill
Lake Farm
Okehampton
Stoke Hill
Exeter
Axminster
Hamworthy
Nanstallon
Fig. 1.

an assumed Roman military ditch between the interim and final reports.[28] In another thirty years the picture will no doubt have changed again just as radically. New sites will be found; old ones may be disproved. Overall the number will certainly increase, giving a picture of an apparently even denser military occupation.

However it is clear, even on our present very imperfect evidence, that not all of these sites were occupied at once. Gloucester, for example, was a late foundation, its construction dated by a coin in a primary foundation trench to no earlier than the late 60s, by which time Hod Hill, Chichester, Lake Farm and Waddon Hill had all almost certainly been abandoned. We will, in fact, never know for certain how many of these military sites were in contemporary occupation, because of the problems of archaeological dating evidence, giving a time span rather than a precise calendar date (hence the severe problems in trying to marry archaeology with history). Moreover many of the squares on the map represent sites which have either not been excavated at all, or have been only very sparsely trenched; an indication of the date of some of these (Cullompton for example) may be obtained from material collected from the surface, but otherwise such sites are included under the designation 'pre-Flavian' on the general historical premise that turf and timber forts are singularly unlikely to have been constructed in this area in the Flavian period (or later) when the military effort in Britain was being concentrated in Wales, in northern England and in Scotland, and the military bases were therefore moved to nearer the scene of the action. If we take, for example, all the sites thought, on the evidence of excavation or surface finds, to have been occupied in the period 55/60 we have a total of about sixteen. Assuming these sites to be more or less densely filled with buildings (an unproven assumption), they would, between them, provide accommodation for well over 20,000 soldiers, that is more than half of the whole campaign force in Britain—a patent nonesense. Not only do we have a very incomplete picture of the internal arrangements of these sites, and hence of their barrack capacity, we cannot date them precisely: a pottery date-span of, say, 50 to 70 may, in reality, reflect occupation for only a brief period in the late 50s and early 60s. Equally, all sites in existence at a given time do not have to be in contemporary occupation. The practice of the modern British army in retaining bases which have not been fully or permanently garrisoned for many years (the Topsham Road barracks in Exeter is a case in point) is one which appears to be paralleled in the Roman army. The refurbishment of the defences of the fort of Hod Hill and the complete reconstruction of the west gateway at Tiverton, may both be due to the need to recommission the site after a brief abandonment, for in both cases the work occurs on a site with a relatively brief overall period of occupation and in neither case does it appear (on the evidence available) to be connected with a change of garrison. Many of the military sites in the South West exhibit more than one structural phase: the two cases quoted represent rebuilding within an unchanged overall defensive circuit, as does Waddon Hill where a tented encampment was replaced by timber buildings: but at Cullompton there appears (on surface observation) to have been a change in the fort dimensions, while at North Tawton there are two completely separate installations (in addition to a temporary camp). At Cirencester, not only are two different units attested (and the site was probably not large enough at any stage of its existence to have accommodated them both together), but there are three phases of structural activity, with a rebuild within a different defensive circuit between phases one and two.[29] The task of building the turf and timber forts characteristic of the early Principate was not a particularly major or time-consuming undertaking: Tacitus records, for example, how the army of Caesennius Paetus on the Euphrates constructed its winter base at the start of the summer season, prior to going off on campaign, and in relation to Agricola's third season in Britain he notes that there was even time for the construction of forts, and that within a summer of active campaigning against new tribes.[30] The crude time-spans provided by archaeological dating evidence will never be adequate to reflect the subtleties of a fluid campaigning situation, when new forts will be built and abandoned, refurbished and reoccupied in response to an ever-changing tactical situation

A further problem which requires consideration, above and beyond the simple existence of these sites, is the question of their nature and their function. The term 'fort' is in one sense a very imprecise word. It denotes, in essence, a defended enclosure providing a home base for soldiers stationed within it, but the reasons for any particular group of soldiers being based in any particular place at any one time are many and varied. The popular picture of the Roman army steadily advancing and dropping down forts in its rear to hold down the newly conquered provincials and hold open the line of retreat, is not a convincing one. Its inescapable corollary is an ever depleting army at the cutting edge. To fragment the army by leaving small scattered garrisons in the rear would be to court disaster. Strength lay in numbers, and if trouble arose the army would fight its way out as it had fought its way in. As Caesennius Paetus observed, the army's strength was not in its fortifications but in its men and its arms: hence its ability and willingness to move freely and speedily over remarkable distances as exhibited by Caesar in Gaul, for example, or, closer in date to Claudio-Neronian Britain, by Corbulo in Armenia. And it is notable that Corbulo, after the defeat of Artaxata preferred to destroy the city than to spare the troops to garrison it: in the midst of a campaign he feared to deplete his army.[31] In the context of active campaigning (as opposed to the rather later static frontier dispositions) the majority of forts will have lain empty during the summer season when the army was engaged in fighting. They will have been occupied only at the close of the campaigning season when the army was distributed *in hibernis*, into its winter quarters, to rest and recoup ready for the forthcoming year. There are. of course, exceptions to this generalisation. Specialised bases existed: hospital bases, for example, supply bases, maintenance depots. Troops might be posted to guard harbours. But all of these would be back in the area already conquered where they would be reasonably secure. In the campaigning area itself forts (the words used in the contemporary literature are *castella* or *praesidia*) might be used tactically, to close off passes, to guard water-supplies, to keep an eye on a native encampment, but these were essentially very short-lived bases and it is doubtful if they contained within them any 'permanent' structures. The troops who occupied them will have been housed in the tents characteristic of the campaigning season. Fragmentation of the army into more permanent bases represents very much a secondary phase, and the distribution of such sites need have no necessary connection with the maintenance of security within their immediate area. Clearly some winter camps may have been maintaining a watch over suspect populations, but others will have been sited with different considerations in mind: the availability of food supplies, for example, or the need to provide protection for an ally. The assumption that an allied kingdom, being a friendly area, will be devoid of a Roman military presence is an ill-founded one. Literature provides many examples of garrisons established within allied territory, while archaeology proves them to have existed within the area under the jurisdiction of Cogidubnus, for example at Fishbourne and Chichester (below). Also within the kingdom of Cogidubnus is the Silchester *oppidum*, and although there is no structural evidence for a fort here, a military presence has long been hypothesised on the basis of a not insignificant collection of early military metalwork.[32] A fort probably also lay at Winchester, adjacent to the *oppidum* of Oram's Arbour, again within the assumed domain of Cogidubnus.[33] Within the area of the allied Dobunnic kingdom lies the Cirencester fort, some 4.5km to the south-east of Bagendon, the major Dobunnic *oppidum*, while a Roman military presence has been suggested at Rodborough Common, 17km to the west of Cirencester, within the Minchinhampton complex, where rescue excavations produced a short stretch of V-shaped ditch, a little military equipment and distinctive pottery types.[34] It is entirely uncertain how long the Dobunnic kingdom kingdom survived, the only allusion to its existence, by name, being in Dio's account of the events of 43; however, in his account of the events of 47 Tacitus records troop movements occasioned by attacks made upon Rome's allies, and given the general context of the passage it is not improbable that the Dobunni are the allies in question. Hence Gloucester-Kingsholm, too, founded in the late 40s, most probably as part of the preparations to launch an attack upon the Silures, may also be construed as lying in allied rather than provincial territory, providing a secure base from which to operate against one of the more troublesome British tribes.[35]

Similar factors apply in the case of the military activity on Chichester harbour. Parts of three Roman military-type granaries have been identified over an area of approximately 0.1ha immediately to the east of the later villas at Fishbourne, whence a creek gave access to the open water. The full extent of this military activity is not known, granaries alone having been identified within such confined areas as have been subject to excavation, but this evidence alone suggests the presence here of a supply base where grain and other perishable foodstuffs shipped in from the Continent or gathered from the surrounding terrain, could be concentrated prior to distribution to the campaigning army.[36] Only 2km west of Fishbourne lies Chichester itself, the modern, medieval and Roman city being set within a broad region protected by a system of dykes, suggesting the presence of a late Iron Age *oppidum*.[37] A Roman military base lay below the later city site, but what relationship this bears to the foci of activity within the *oppidum* is unknown since minimal evidence of pre-Roman activity has as yet been picked up within the built-up area. The Roman military activity is known to extend over an area in excess of 16ha and may well be more extensive than this since the defences have as yet been located on the eastern side only.[38] Another probable harbour site of military origin lies some 90 km to the west at Hamworthy on Poole Bay. Unfortunately this site disappeared into a quarry and under housing development without systematic excavation ever having taken place, but early Roman pottery and coins suggest a military connection for this very likely harbour site.[39] Moreover, Hamworthy is linked by road to the base at Lake Farm where Roman military activity extends over an area of at least 13ha. On the evidence of the most recent excavations, there appears to be essentially one phase of occupation here, confined within the period c.45-60/65AD.[41] The configuration of the south-western peninsula invites the systematic use of water transport for supply purposes, and other coastal storage facilities with associated military base are doubtless to be found. The Weymouth area, with its known Roman road link northwards to Dorchester is a distinct possibility, as is the region of the Axe mouth. The context for the Claudio-Neronian material at Topsham, once thought to be military, is now known to be civil; but a supply-base here, mid-way between Exeter and the sea, is not really necessary, given that the Exe may well have been navigable as far as the fortress itself. A quay very probably lay on the east bank of the river below the military site.[41] Further west, Mount Batten, an entrepot of pre- and later Roman date, is an attractive possibility. On the north Somerset coast Sea Mills has produced Claudian pottery and coins, though structural evidence of appropriate date is still lacking, while an early military quay may be assumed on the Severn at Gloucester.[42]

With the exception of the two (relatively late) legionary bases at Exeter and Gloucester, the majority of the south-western fort sites are small in size, capable of accommodating no more than about a thousand men, frequently less. The exceptions to this generalisation include three of the earliest sites within the region, that is Chichester, Lake Farm and Gloucester-Kingsholm, none of certain size, all apparently within the range 13 to 18ha. Another exception lies at North Tawton. Here a third military-looking enclosure has recently been identified from the air, lying between the previously known fort and camp.[43] Only part of the defensive circuit has as yet shown: its north-west corner and a length of double ditch on its north side, together with the position of a probable northern entrance, suggest that it will originally have been a minimum of 5ha in area (assuming a 'normal' length breadth ration of 3:2). However, superficially the closest parallel to North Tawton is the category of site commonly known as 'vexillation fortress', and best exemplified by Longthorpe.[44] Such information as we have on these 'vexillation fortresses' suggests that they belong to the early phases of the conquest, and this would be consistent with the evidence at North Tawton where another, considerably smaller (and putatively later) fort exists on a completely separate site to the south. It should however be emphasized that this view of the North Tawton sequence is essentially speculative: we have no positive dating evidence for the large site and a later Roman date cannot be entirely ruled out.[45] Whether or not one accepts a first-century date for the entire North Tawton complex, the site is clearly a critical one, placed at a crossroads where the main east-west route down the peninsula is joined by a road through the Taw-Torridge watershed, leading towards Alverdiscott and the north Devon coast. As yet no further large early bases have been positively identified in the South West. The case for a large early phase at Cirencester has now been reviewed and recanted, though speculation has arisen about the existence of a substantial military site at Ilchester.[46] In discussion of 'vexillation fortresses' it is normally assumed that they held a detachment of legionaries plus a number of auxiliaries, often cavalry to complement the infantry strength of the legion. This particular combination, though likely, is unproved so no attempt is being made in the present context to speculate upon the nature of the garrison within these south-western sites, except to recall the recovery from the vicinity of the Gloucester-Kingsholm site of two tombstones, one legionary and one cavalry (above 3, 4-5). Whether these other large sites contained all legionaries, legionaries plus auxiliaries or a grouping of auxiliary units is unknown.[47] However, by about 60, and perhaps as early as 55, a base had been established for legion *II Augusta* at Exeter, which retained a military presence until the mid-70s. Exactly how long *II Augusta* remained in occupation has been a matter of debate, tied up with the problem of the garrison of the Gloucester site. No earlier than AD 66 a 17.5-ha legionary fortress was constructed a short distance to the south of the previous base at Kingsholm. Until very recently it has generally been accepted that legion *XX*, formerly based in the west, had moved north to Wroxeter in or about AD 66 to replace legion *XIV* which Nero had moved out of Britain in that year as part of troop movements connected with his projected eastern campaign. It seemed logical to assume that the new Gloucester fortress (its construction dated by coins of 64 and 66 in primary positions), was built to house a new legion, and since *IX* was at this date firmly ensconced at Lincoln, that left only *II Augusta*. However the Exeter evidence, as interpreted by its excavator, did not seem happily to accommodate this hypothesis, for it suggested, on the basis of coin and samian pottery evidence, that *II Augusta* remained in occupation at Exeter until c. 75 when it moved direct to Caerleon, allowing no time for a sojourn at Gloucester.[48] An alternative interpretation of the evidence (advanced before the publication of the Exeter material), was proposed by Hassall and Rhodes who suggested that the Gloucester fortress was built at the time when legion *XIV* was posted briefly back to Britain during the civil war of 69, and the legionary establishment of the province thus restored to four.[49] The brevity of the fourteenth's stay in Britain does however make this rather unlikely, and in any case cannot account for the whole period of the fortresses's occupation. While the coexistence of Exeter and Gloucester of itself provides no problem, their contemporary *occupation* does, and the continued use of the Exeter site through until the mid-70s does seem to be indicated by the coin and samian evidence. Was the legion split? Did one or other site contain a part (or wholly) auxiliary garrison? Or is the overlap more apparent than real, an effect of the elongation of dates produced by archaeological evidence, particularly if the foundation of Gloucester is pushed back by a few years into the early 70s? It is commonly assumed that the shift to Gloucester was an immediate knock-on effect of the move of the XIVth, but it is equally likely that it was occasioned by the decision to resume advance into Wales, a decision not taken, as far as is known, until the accession of Vespasian, and effected during the governorship of Julius Frontinus in AD 73-77. A preparatory shift of base in the early 70s would as well accommodate the archaeological dating evidence as would the assumption of a shift in the late 60s. It would give legion II a brief sojourn at Gloucester prior to its move to Caerleon, a fortress whose internal layout is now known not to have been completed until some decades after its initiation consequent upon the conquest of the Silures by Frontinus.

The recent discovery of a twentieth legion building inscription, reused in Gloucester cathedral, alters but does not solve the problem of the Gloucester garrison.[50] The inscription is a centurial stone, dating to AD 61 or later, and is more likely to derive from a masonry than a timber structure. Hence, unless it comes from an unknown bathhouse in the Kingsholm complex it must relate to the Gloucester-centre fortress which, in its second phase, acquired some stone buildings and possibly a stone defensive wall.[51] Its reuse in the cathedral is consistent with this belief. Hurst therefore suggests that *XX* never moved away to Wroxeter but remained in the west, building and occupying the Gloucester fortress from the late 60s. It then moved away temporarily to an intended base at Inchtuthil in the 80s, returning to Gloucester when the Scottish conquests were abandoned in or soon after AD86. It is now that rebuilding in the fortress produced the first stone structures from which Hurst suggests the twentieth legion inscription derives. Military occupation of the site continued down into the 90s when the legion moved on and the fortress was turned into a *colonia*, its foundation dated, on place-name evidence, to the reign of Nerva, AD 96-98.[52] Legion II is thus freed to remain at Exeter until its move direct to Caerleon in the mid-70s. Hence one problem is solved by this hypothesis; several others are, however, created. Wroxeter, hitherto believed to be the home of legion *XX* from c. 66 until c. 86 is deprived of a garrison for this entire period (except perhaps for the few months when *XIV* was back in the province); likewise Chester, where *XX* was thought to have moved in the late 80s (replacing *II Adiutrix* when it was transferred to the Danube), will have lain empty for almost a decade, and all this at a time when, on Hurst's reasoning, two legions lay close together in the west, first at Exeter and Gloucester, then at Gloucester and Caerleon. It is hard to conceive of the continuing isolation of a legion at Exeter at a time when Wroxeter lay empty, and the Welsh marchland bereft of a legion at critical periods during the conquest of Wales.

This scenario is not a compelling one, but can be avoided if the inscription is assumed to derive from a civil rather than a military phase of the city-centre site. Legions were commonly involved in civic construction work, specifically (in the present context) in building work on veteran colonies, and Gloucester was a colony for veterans of the XXth legion.[53]

Given the small number of legions within any individual province at any one time the siting of their bases tends to be determined less by local than by much wider strategic considerations, hence the importance of the strategic road network which allowed them to deploy swiftly around the province. Rather different factors apply in the case of the smaller forts which form the bulk of the known military sites in the area. These vary in size from less than one hectare (Nanstallon) to slightly over three (Shapwick and North Tawton), with the majority less than two. Hod Hill, an apparent exception at 4ha, does in fact fall into this same category since its eccentric shape, caused by the fact that it was inserted into the corner of an Iron Age hillfort, means that an area of only 1.8ha was actually occupied by buildings (the estimated equivalent figure for North Tawton is 1.9). The sites at the larger end of this range will have had a capacity of anything up to 1500 men, or rather less if cavalry were involved, the smaller ones no more than 500. The bulk of them lie along known or assumed road lines, suggesting either their survival through into a period when the provincial infrastructure was being laid out, or their subsequent foundation. One interesting group of sites which does not conform to this norm are those which occupy hill-top positions, taking over the sites of Iron Age hill forts. Hod Hill falls into this category, as does Hembury and South Cadbury and (assuming a Roman fort did exist there) Ham Hill. Todd has also suggested that a military presence may have been briefly maintained at Maiden Castle, thus explaining the presence of mid-first century Roman pottery on the site.[54] The use of elevated, relatively inaccessible sites of this type, while not conforming to the criteria of 'typical' fort siting, which dictate accessibility, ease of movement, ease of access to a water-supply, is not uncommon in forts founded apparently in the immediate wake of a campaign: where datable they tend to represent an early strand within the military occupation of an area. While they come together as a group because they share a common distinctive topographical factor in terms of their general siting, the detailed reason for choice of each individual site will have varied from place to place. The fort at Hod Hill, for example, appears to have been imposed on a site that had just been conquered, in contrast to Hembury where the latest Iron Age phases are missing. The Roman army here apparently chose a dominant but abandoned site as a secure base with a wide outlook over the countryside around.

The relationship of fort-siting in general to the local settlement pattern is a matter of considerable interest and importance, but one on which it is at present possible to do little more than speculate, for so very few Iron Age sites in the area have been investigated, and even fewer shown to have been in existence on the eve of the Conquest. It is often observed that such and such a Roman fort lies adjacent to such and such an Iron Age hillfort, and it tends to be assumed, if not explicitly stated, that the two are connected, militarily speaking. A comparison of fort and hillfort distributions in the South West indicates that the majority of Roman military sites lay in general proximity to a native fortification, though rarely closer than a couple of kilometres (Fig.2). Given the multiplicity of fortified hills and promontories in the region it may questioned whether such juxtapositions are any more than random. In some at least of these cases the native site will have been abandoned prior to the Roman arrival: the case of Hembury has already been quoted; St Catherine's Hill, Winchester, represents another example. Halstock, Okehampton, is undated, as is Cranmore Camp, Tiverton, King's Castle, Wiveliscombe, and the hill-slope enclosure immediately adjacent to the Alverdiscott camp. While all may be taken as being broadly Iron Age in date there is no guarantee that any or all of these sites were still in occupation in the 40s and 50s when their adjacent Roman forts were built. Contemporaneity can rarely be actually demonstrated, and while it could be argued that the siting of a fort in the vicinity even of an abandoned site would deny its use to a potential dissident, clearly all potential refuges could not be thus guarded except by general observation and patrol which requires not a close single fort/hillfort relationship, but the distribution of troops generally over the area so as to give good overall coverage. Hence one of the factors leading to the development of fort networks, the fairly regular spacing of sites, about twenty or so kilometres apart, linked to one another by roads. However, as noted above, such a scattering (and hence weakening) of manpower would not be anticipated until the initial battles had been fought and won, the back of the opposition broken. The maintenance of troops in appropriately large battle-groups, essential during the initial campaigning, did not, however, facilitate the police-work which followed: this led to the distribution of troops into scattered forts from where they might be broken down into even smaller units (a century or less in some fortlets) for specific jobs of observation and control.[55] Another factor which will have favoured the scattering of troops once the major threat was past, will have been the problem of supply, distributing the burden of provisioning the army at a period when it was no longer necessary (or desirable) for all foodstuffs to be shipped in from the Continent or stripped off the land as would happen in wartime. The problem of land pollution, too, may have been a not inconsiderable factor in the decision to break up major troop concentrations.

Having once decided to scatter the troops, the detail of fort siting would be determined in large part by topographical considerations; by the

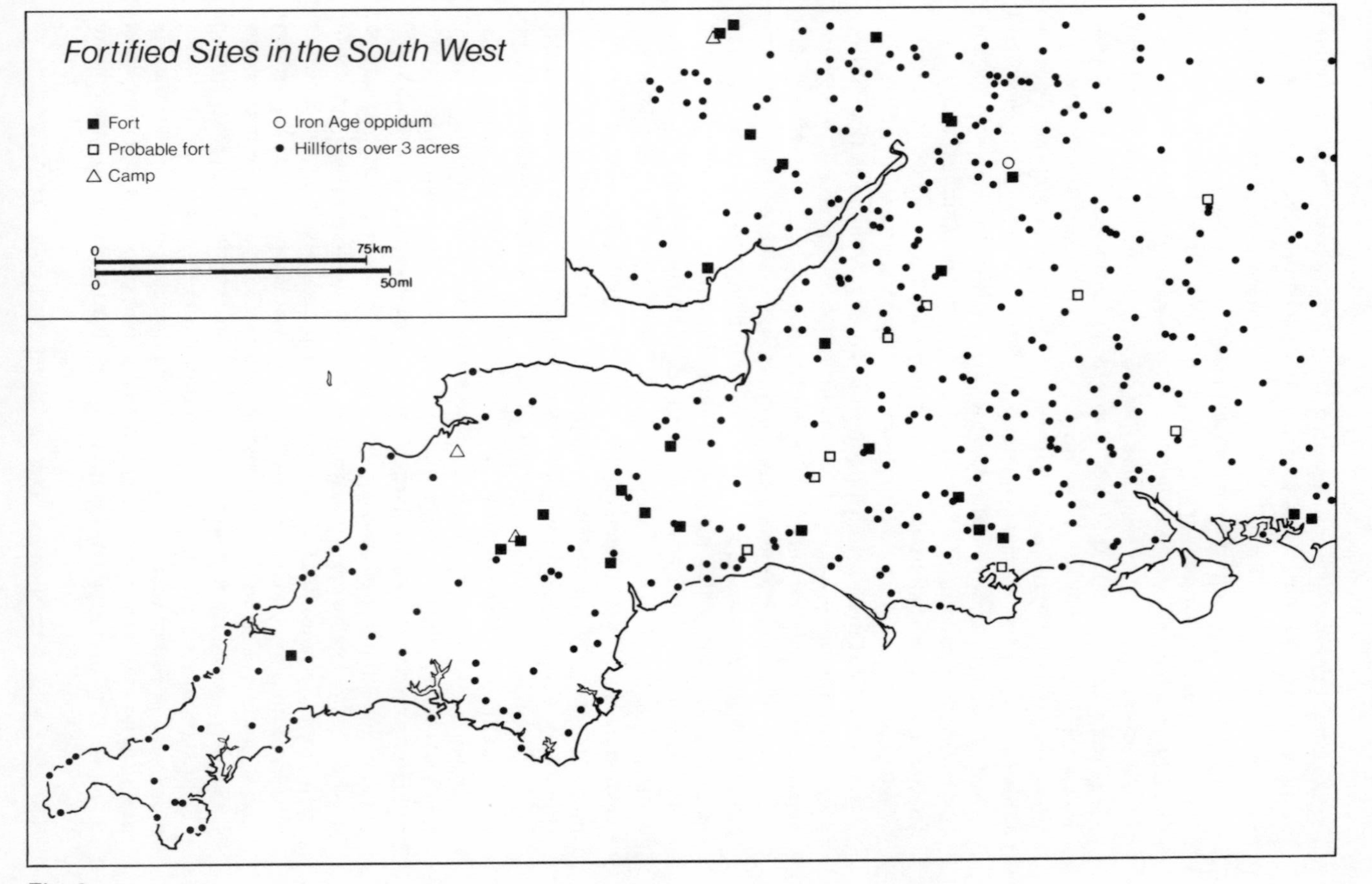
Fortified Sites in the South West
Fort
Probable fort
Camp
Iron Age oppidum
Hillforts over 3 acres
0
75km
0
50ml

Fig. 2.

availability of water for drinking, but combined with a slightly elevated well-drained site so that the readily-available water did not constitute a flood problem; by the need for ease of access, hence on natural routeways, leading often to bridge-head positions overlooking river crossings. Other factors might apply in specific cases, as, for example, the stationing of troops to guard and supervise the exploitation of mineral resources. At Charterhouse-on-Mendip lead/silver extraction had begun by AD 49 and military control is attested by the pigs bearing the stamp of legion II as well as by the presence of a fort.[56] Nanstallon, near Bodmin, is sited in an area rich in minerals, a factor which may have rather more to do with its siting than any supposed disaffection in the South West at the time of the Boudican rebellion.[57]

A military presence was maintained in the South West until the mid-70s, either because it was needed here or (more likely) through inertia, because the troops were not needed elsewhere. During the 70s and 80s however, direct Roman control was extended into Wales, northern England and Scotland and the army was moved out of the long-pacified west country (as out of the rest of south and east England) so as to bring it closer into touch with these new areas of military importance.

The legacy left behind was two-fold; the road structure and the fort sites, certain of which were chosen as administrative foci. The *civitates* of the South West (in common with those of the rest of Britain), were probably all centred on former fort sites, though this has yet to be positively demonstrated at Dorchester (Dorset) and Winchester. The reasons for this are partly topographical (the factors which make for a good fort site are also suitable for a city site), and partly linked with the administrative structure which was already perhaps starting to emerge during the period of military control.

Once the army had abandoned its south-western bases, soldiers will rarely have been seen en masse. They will have come from time to time as escorts to officials, in connection with tax collection for example; in a police capacity in connection with civil jurisdiction; collecting supplies, meat, sheepskins, leather or pottery, for example—pottery from the Corfe Mullen kilns in Dorset has been identified on military sites as far north and west as the army went. Veterans will have retired to the colony at Gloucester and new legionaries will no doubt have been recruited there. Meanwhile the non-Roman communities, the peregrine *civitates* provided recruits for the auxilia.[58] Resistance to Rome put down, internal security assured, the Roman army was distanced from the South West. Not until the late Roman period might the army have returned as a fighting force, and then not as an aggressor; the erstwhile enemy was needed to maintain the integrity of the province from the next wave of external aggressors.

NOTES

1. Cassius Dio, *Roman History* LX, 19-22.
2. Tacitus *Annals*, 12.31.
3. Suetonius, Vespasianus 4. The reference to the command of the second legion appears in Tacitus Histories 3.44.
4. Birley, A., *The Fasti of Roman Britain* (1980), 225–8.
5. Ptolemy, Geography ii.3.13. The sources used by Ptolemy are discussed by Rivet, A. & Smith, C., *The Place-Names of Roman Britain* (1979), 114-15.
6. Bidwell, P. & Boon, G., 'An antefix of the second Augustan legion from Exeter', *Britannia* 7 (1976), 278-80.
7. Tacitus, *Annals*, 12.32; Collingwood, R.G. & Wright, R.P, *The Roman Inscriptions of Britain* (1965)(hereinafter *RIB*), 122.
8. Hurst, H.R., *Kingsholm. Gloucester Archaeological Report* 1 (1985).
9. Manning, W., *Report on the Excavations at Usk 1965-1976. The Fortress Excavations 1968-71* (1981), 38-39, pl. 1.
10. CIL XVI 69, records 37 cohorts and 13 alae.
11. Holder, P. *The Roman Army in Britain* (1982), 107-124.
12. The earliest dates to AD 98: CIL XVI 43.
13. *RIB* 108, 109 (Cirencester); *RIB* 201 (Colchester).
14. Hassall, M, 'Epigraphic evidence for the auxiliary garrison at Cirencester', in Wacher, J. & McWhirr, A., *Early Roman Occupation at Cirencester* (1982), 67-71.
15. *RIB* 121. The discovery of a cheek-piece from a cavalry helmet need have no relevance for the garrison of Kingsholm, since the piece was only partially fabricated and could therefore be the product of a workshop at Kingsholm supplying equipment to sites in the neighbourhood. Moreover, it is unclear from which phase of the site the piece derives, it being attributed variously to phase 1 (Hurst o.c. note 8, pages 6 & 117) and to phase 2 (page 122).
16. *RIB* 159 (Bath); *RIB* 403 (Brecon Gaer).
17. CIL VI 3539 = ILS 2730; Maxfield, V.A., *The Military Decorations of the Roman Army* (1981), 163-64.
18. Holder, P., *The Auxilia from Augustus to Trajan* (1980), 30-31; *The Roman Army in Britain* (1982), 22, 110.
19. CIL XVI 26; Tacitus, *Agricola* 32; AE 1956, 249; CIL XVI 49, 163. Recruitment of Britons to the Roman army is discussed by Dobson, B. & Mann, J.C, 'The Roman army in Britain and Britons in the Roman Army', *Britannia* 4 (1973), 191-205.
20. *RIB* 92 with amendment by Bogaers, J., 'King Cogidubnus: another reading of *RIB* 92', *Britannia* 10 (1979), 243-254.

21. Alcock, L., 'Excavations at South Cadbury Castle 1970: a summary report', *Antiq. J.* 51 (1971), 1-7 suggests a Claudian date; Alcock, L., *By South Cadbury: is that Camelot?* (1972), 159-72 an early Flavian date; Manning, W., 'The conquest of the West Country', in Branigan, K. & Fowler, P., *The Roman West Country* (1976), 37-38 suggests a context associated with the Boudiccan rebellion of 60-61.

22. Dymond, C.W., *Worlebury* (1902).

23. Burrow, I., *Hillfort and Hilltop Settlement in Somerset in the First to Eighth Centuries AD* (1981) takes the massacre to be Roman: Harding, D., *The Iron Age in Lowland Britain* (1974), 222 opts for an Iron Age context; Cunliffe, B., *Iron Age communities in Britain* (1975), 102 appears to favour the later Iron Age, though with some equivocation.

24. Cunliffe, B., 'Danebury, Hampshire: First interim report on the excavation 1969-70', *Antiq. J.* 51 (1971), 246, 251; Dyer, J., *Southern England: an Archaeological Guide* (1973), 154 discards the reservations. Final report, Cunliffe, B., *Danebury, an Iron Age hillfort in Hampshire. Vol. 1 The Excavations 1969-1978: the Site* (1984), 45-46.

25. Cunliffe, B., *Iron Age Communities in Britain* (1975), 260-63.

26. Todd, M., 'Excavations at Hembury, Devon, 1980-83; a summary report', *Antiq. J.* 64 (1984), 251-268.

27. Threipland, L.M., 'An excavation at St. Mawgan-in-Pyder, North Cornwall', *Archaeol. J.* 113 (1956), 33-78 esp 52; Fox, A., *South West England* (1973), 149, 165.

28. Wedlake, W., *The Excavation of the Shrine of Apollo at Nettleton Shrub, Wiltshire, 1956-1971* (1982), 7. The existence of a fort somewhere in the vicinity cannot however be excluded, given the presence at Nettleton of military metalwork, a significant quantity of Claudian coinage and pottery characteristic of mid-first century military sites.

29. Wacher, J. & McWhirr, A., *Early Roman Occupation at Cirencester* (1982).

30. Tacitus, *Annals* 15.7; *Agricola* 22.

31. Tacitus, *Annals* 15.10; 13.14.

32. Boon, G., *Silchester: the Roman town of Calleva* (1974), 47 & fig.8.

33. Biddle, M., 'Excavations at Winchester 1971. Tenth and final interim report: Part II', *Antiq. J.* 55 (1975), 295-337 esp 295-97. The reason for these forts is only conjectural, but some unrest within the kingdom of Cogidubnus would not be unexpected, given that he had been appointed ruler over an amalgum of lands; Tacitus , *Agricola* 14.

34. Swan, V.G., 'Oare reconsidered and the origins of Savernake Ware in Wiltshire', *Britannia* 6 (1975), 37-61 esp. 4. The Bagendon and Minchinhampton earthworks are discussed in Clifford, E.M., *Bagendon. A Belgic Oppidum. Excavations 1954-1956* (1961).

35. Tacitus, *Annals* 12.31 records the attack on Rome's allies. For the foundation of Gloucester-Kingsholm see above page 3.

36. Cunliffe, B., *Excavations at Fishbourne. Vol. 1 The Site* (1971), 37-42. Recent rescue excavations at Fishbourne (*Britannia* 15 (1984),328) have also produced evidence for part of a granary.

37. Bradley, R., 'A field survey of the Chichester Entrenchments', in Cunliffe, B. op. cit. note 36, 17-34.

38. Down, A., *Chichester Excavations 3*, Fig. 4.1, 4.2.

39. *Proc. Dorset Natur. Hist and Archaeol. Soc.* 52 (1930), 96; 56 (1934), 11.

40. Information from the excavator Ian Horsey, Poole Museums. Earlier limited excavation on the site had suggested it to be multi-phase.

41. Cleere is among those who accepts, without question, the existance of a Roman port at Topsham: 'Roman harbours in Britain south of Hadrian's Wall', in Taylor, J.du Plat & Cleere, H., (eds), *Roman Shipping and Trade* (1978), 36-40. For a review of the Topsham evidence in the light of recent excavation, see Maxfield, V.A., 'The Roman Occupation of South-West England: further light and fresh problems', in Hanson, W. & Keppie, L.J.F. (eds), *Roman Frontier Studies 1979* (1980), 297-309, esp. Appendix. 305-307.

42. Boon, G., 'The Roman Site at Sea Mills 1945-6', *Trans. Bristol and Gloucs. Archaeol. Soc.* 66 (1945), 258-95; 'A Claudian origin for Sea Mills', *Trans. Bristol and Gloucs. Archaeol. Soc.* 68 (1949), 184-88.

43. Griffith, F., 'Roman military sites in Devon: some recent discoveries', *Proc. Devon Archaeol. Soc.* 42 (1984), 20-24, figs 4 & 5, pl. 6.

44. If the extant north gate of North Tawton were centrally positioned, the east-west dimension of the fort, measuring to the outer lips of the outer ditch, would be just over 380m; this compares with the 374 of Longthorpe, an almost square site with an area (over the ramparts), of 10.9ha. Frere, S. & St.Joseph, J.K., 'The Roman fortress at Longthorpe', *Britannia* 5 (1974), 1-129.

45. Todd has suggested (in verbal discussion of this paper) that the large enclosure is not a first century fort but a late Roman statio, drawing attention to the situation at Bury Barton where a 1.9ha fort of mid-first century date, lies within a larger earthwork initially assigned to an early Roman context but which he has now redated to the late Empire: Todd, M., 'The Roman Fort at Bury Barton, Devonshire', *Britannia* 16 (1985), 49-56. While the question will be resolved only by fieldwalking or excavation, the superficial resemblance of the North Tawton site to the large double-ditched forts of the mid-first century date is striking.

46. Wacher, J. & McWhirr, A., *Early Roman Occupation at Cirencester* (1982). Information on Ilchester from John Casey.

47. On the problem of identifying troops on the basis of archaeological evidence cf. Maxfield, V.A., 'Pre-Flavian forts and their garrisons', *Britannia* 17 (1986), 59-72.

48. Bidwell, P., *The Legionary Bathhouse and Basilica and Forum at Exeter* (1979), 16; Bidwell, P. & Boon, G., 'An antefix of the Second Augustan Legion from Exeter', *Britannia* 7 (1976), 278-280.

49. Hassall, M. & Rhodes, D., 'Excavations at the new Market Hall Gloucester, 1966-7', *Trans. Bristol and Gloucs. Archaeol. Soc.* 93 (1974), 31.

50. Hassall, M.W.C. & Tomlin, R.S.O., 'Roman Britain in 1985: Inscriptions', *Britannia* 17 (1986), 429.

51. Hurst, H.R., *Gloucester, the Roman and Later Defences* (1986), 118-19.

52. CIL VI 3346; Rivet, A.L.F., & Smith, C., *The Placenames of Roman Britain* (1979), 368.

53. CIL VIII 17842-43 record legion *III Augusta* doing building work at its colony at Timgad in North Africa. References to soldiers engaged in civil building projects in general are collected by R.W.Davies in 'The Daily Life of the Roman soldier under

the Principate', in Temporini, H. (ed.) *Aufstieg und Niedergang der römischen Welt II.1* (1974), 329-331.

54. Todd, M., 'The early Roman phase at Maiden Castle', *Britannia* 15 (1984), 254-55.

55. For example, the fortlets of Old Burrow and Martinhoe on the north Devon coast, look out towards the south Wales coast, and were probably connected, as their excavators argued, with the maintenance of security agaist the (as yet unconquered) Silures. Fox. A. & Ravenhill, W., 'Early Roman outposts on the north Devon coast, Old Burrow and Martinhoe', *Proc. Devon Archaeol. Soc.* 24 (1966), 3-39.

56. Elkington, H.D.H., 'The Mendip lead industry', in Branigan, K. & Fowler, P. (eds)., *The Romans in the West Country* (1976), 183-197.

57. Fox, A. & Ravenhill, W., 'The Roman fort at Nanstallon, Cornwall', *Britannia* 3 (1972), 56-111. Lead, silver, copper and gold were all locally available as well as iron ore.

58. Op. cit. note 19.

Public and Private Defence in the Medieval South West: Town, Castle and Fort

BY ROBERT HIGHAM

This essay discusses problems of defence and security from the end of Roman provincial history to the end of the fifteenth century. Its theme is the response which various societies and individuals made to threats from outside the region and from within it. The emphasis is on fortifications rather than the raising and organization of armies, but the purpose is not to present the physical and documentary evidence for all fortifications built in this long period. It is rather to examine the changing character of defensive sites and the changing circumstances which gave rise to them. Selected examples are drawn from the four south-western counties, but this is not the place for a general history of fortifications in this large region. Places mentioned are shown on the accompanying map (Fig. 3).

The approach is essentially secular, and the defence of ecclesiastical interests is largely neglected. The latter is an important subject worthy of its own treatment. The foundation charter of Exeter Cathedral (1050) mentioned that an advantage in merging the dioceses of Devon and Cornwall was to provide better protection against pirates. Medieval bishops enclosed their cathedral precincts and adjacent palaces with walls (for example, Exeter and Wells), received licences to crenellate for other properties (for example, Chudleigh, Devon) and indeed built traditional castles in earlier times (for example, Old Sherborne). Equally the subject of prisons is not dealt with here, though they were one response to problems of security. Sometimes a prison was found within a castle, as at Exeter. The most famous is probably Lydford, which acted as the Stannary prison. But in the middle ages prisons were not necessarily custom-built. Robert of Bellême

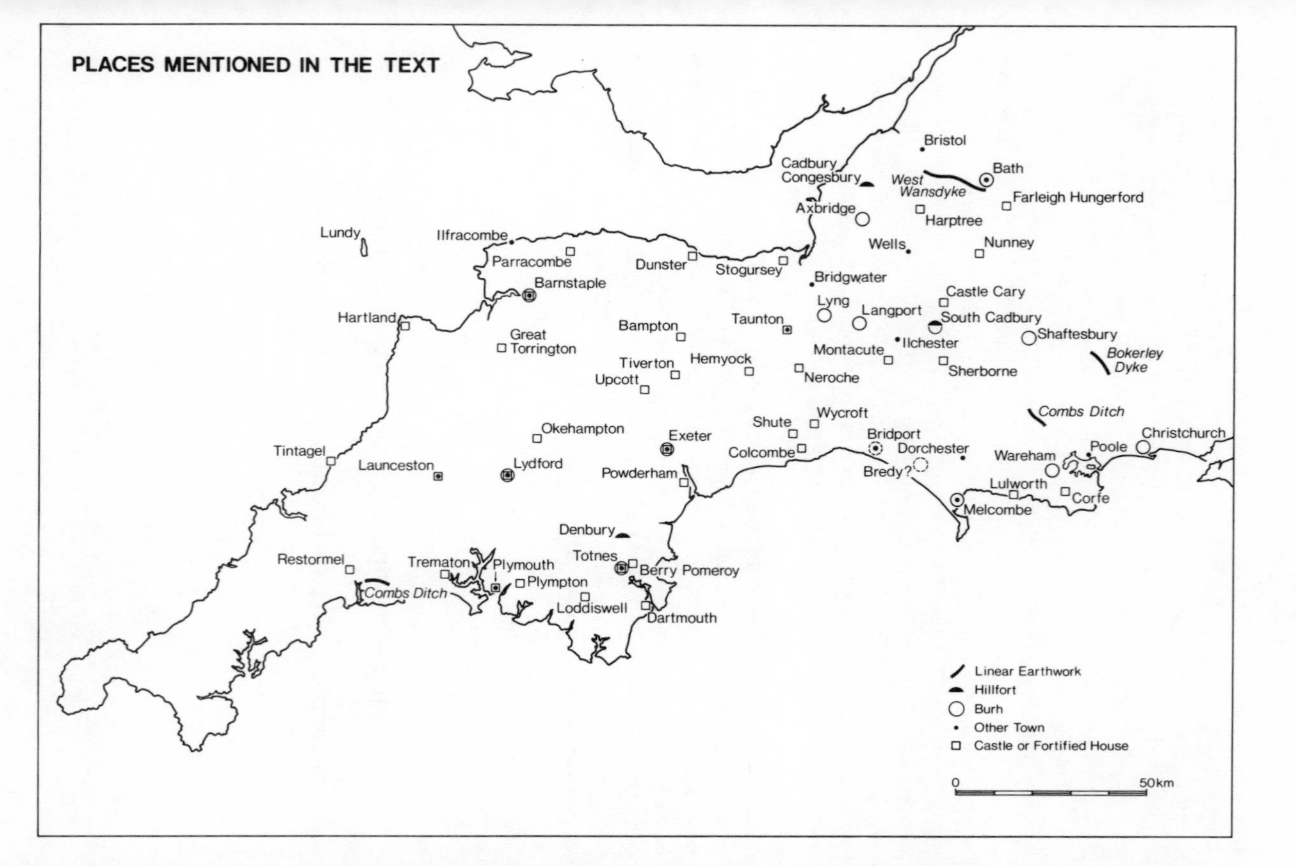

Fig. 3. Map of places mentioned in the text: fortified sites of the Middle Ages. It is difficult to give all the evidence for late Saxon towns in simple form. Some Burghal Hidage names are not shown separately, viz. Pilton (near Barnstaple), Halwell (near Totnes), *Hlidan* (Lydford), Twinham (Christchurch). Towns not named in the Burghal Hidage but nevertheless existing include Dorchester and Ilchester. Alternative identifications for the *burh* at Bredy are given.

spent his last years incarcerated at Henry I's insistence in Wareham castle. A final theme deserving its own study on another occasion is that of the South West as a *march*. The western parts of the region would bear useful comparison with other English extremities such as the far north or the Welsh border. Cornwall in particular was absorbed into the West Saxon kingdom very late, whereas the eastern end of our region contained some of the heartland of Wessex. After the Conquest, although Robert of Mortain was not called 'earl of Cornwall', his great concentration of landed wealth in that county suggests the Normans viewed the area in the same way as other border territories, an attitude anticipating the later earldom and Duchy.

Who built defences and on whose behalf?[1] The Celtic leaders who defended the diminishing kingdom of Dumnonia had a dual notion of their leadership. On the one hand they inherited the responsibility of late Roman government for the protection of the province and the civilized life enjoyed there, against barbarian threat. On the other hand, the development of new Celtic dynasties encouraged a more personal aspect of leadership, with a consequently greater emphasis on the function of protecting people rather than places. In the following centuries the notion of kingship among the West Saxons shared a Germanic and Roman ancestry with kingship on the continent. It had a strong characteristic of protective responsibility, stemming from the view that the authority was god-given. According to this theocratic style of kingship, it was a duty of kings to protect their people and their kingdom, and keep an orderly society. The *burhs* of the Saxon South West should be seen in this light, and their discussion will centre round their fundamentally royal character. Another aspect of early kingship was the protection of royal rights. Throughout the middle ages the building of fortifications was regarded as a royal right, which could be delegated at the king's discretion but not assumed by others against his will.[2] In the post-Conquest period this notion was applied to the growing practice of building private castles. Both in England and on the continent the theory was that they were ultimately all royal (or in some areas ducal) castles, a belief whose origins can be traced back into the Carolingian period. In this sense the customary distinction between the Anglo-Saxon *burh*, a supposedly communal defence, and the Norman castle, a supposedly private one, may be somewhat misleading, since both had their origin in royal authority and both were one means by which the royal duty of protection of individuals and society at large was carried out. The difference was that whereas the *burhs* remained in immediate royal control, castles became widely dispersed (and not always in the king's interest) throughout the upper levels of society, a reflection of the more devolved and contractual nature of feudal government, as well as of the failure of kings always to apply the theory of royal control, to protect this particular regalian right from alienation. In England, of course, it also reflected the take-over of a new landholding class after the Conquest and its need to establish itself in a hostile environment. And in specific areas, notably the Welsh border, the authority to build castles was deliberately delegated by the kings. The difference between so-called theocratic and so-called feudal kingship should not be particularly stressed where the matter of protective duties is concerned. The duty of protecting the kingdom and ensuring law and order persisted from the former into the latter, though it was complicated by the growth in political terms of the personal obligation between the king and his vassals. This in itself, of course, was not entirely new, the earlier centuries being well acquainted with the lord-man relationship. Equally, at the end of the period under review, it is important not to exaggerate the change from a 'medieval world' to a 'modern world' somewhere around AD 1500, an idea which pervades a great deal of thinking about fortifications. In terms of political theory the concepts of 'state', 'crown', and 'community of the realm' were emerging from the thirteenth century, and in England in particular more collaborative forms of government developed naturally from the strong tradition of contractual feudal government. Equally, there were many features of medieval government which persisted long after 1500. What affected changing styles of fortification was not so much a general transformation of society as piecemeal developments, over several generations, in the standards of domestic comfort expected within fortified sites, in the technology of warfare and in the nature of the threats which fortifications had to meet. For such reasons both military fortifications and domestic defences evolved between the fourteenth and sixteenth centuries.

In what follows it will become apparent that the traditional classification of medieval defences is too rigid.[3] We are accustomed to thinking in terms of the Anglo-Saxon *burh*, a town whose defences were designed to protect society as a whole; of the Norman and later castle, which in contrast was a private affair designed to protect only its lord and his followers; and of the artillery fortifications, from Tudor times onwards, representing a new concept, both physically and socially, the defences of central government in a recognizably modern form. In fact, the differences are far less clear-cut than this, and it is more useful to think in terms of a gradual flow of developments rather than in hard categories. Second it will emerge that whereas fortifications are normally seen as reactions to external enemies, for much of the period under review it was the threat of disorder within society which was the major concern and which encouraged so many efforts in the building of fortifications. It is too easy for problems of outside threat to dominate discussion of defence, particularly in an area such as the South West with a vulnerable coastline. But in fact the majority of defensible sites in the period as a whole were built for quite different reasons.

Dumnonia and its neighbours

The emergence of the kingdom of Dumnonia out of the vestiges of the late Roman province is an obscure subject. By the mid-sixth century, when the Welsh monk Gildas was writing, it was ruled by a king called Constantine, but the process by which a new order had been created is not known. Equally, despite its apparent coherence, the Anglo-Saxon Chronicle compiled centuries later gives only a selective view of the origin and expansion of Wessex. It was probably a process of piecemeal land-grabbing, punctuated by major offensives. Some of these stayed alive in oral tradition and were eventually written down, such as the battle of Dyrham in 577, dividing the British of the South West from those of Wales, and Egbert's victory over the Cornish at Hingston Down in 838.[4]

We do not know how common the use of fortified sites was in this period, but it may well have been more extensive than the available evidence suggests. This falls into two categories,linear earthworks and hill-forts. Bokerley Dyke in Dorset has been shown by recent fieldwork by the Royal Commission on Historic Monuments to have a much longer ancestry than previously supposed, originating in prehistoric times.[5] Its later history was probably as a late Roman estate boundary, but this was shown by excavation to have been extended first in the late fourth century and again sometime in the fifth century across the Roman road system towards Dorchester, the administrative centre of the Durotriges. It was almost certainly a barrier to Saxon expansion, built by the British inheritors of Roman authority,and in view of the thin evidence of pagan Saxon influence in the area was successful for most of the fifth century. Combs Ditch, fifteen miles further along the same road system, may well have been its successor, representing a retreat of native control.[6] The West Wansdyke, in Somerset, has always presented major problems of interpretation. Like the East Wansdyke in Wiltshire, it was known by this name by the late Saxon period, but it cannot be shown just when it had been built. It may consequently have been a British work built against Germanic settlers to the north, or equally a later West Saxon work built against their Mercian neighbours.[7] Another example of this class, which has not received the attention it deserves, is the Giant's Hedge in south Cornwall. While it cannot be shown to be of this period it may well represent the boundary of a sub-division of Dumnonia or of one of its last remnants in the ninth or tenth centuries.[8] Whatever the precise origins of these monuments, their scale must have been dictated by significant needs. They imply a massive organization of labour, and presumably reflect the decisions of military and political leaders. They are most unlikely to have been products of the co-operative labours of rural society. But they were designed to protect whole territories and societies, and in this respect follow the same tradition as their Roman antecedents and some of their medieval successors. But we know virtually nothing of their organization. Were they garrisoned? Did they have defended gateways, or palisades on their crests? Was the labour which built them pressed especially for the occasion, or was there at this early date a regular obligation on the part of landholders to provide labour for fortifications? This was certainly the case later, in both Saxon England and Celtic Wales. They were part of a tradition whose other famous examples are found in East Anglia and on the Welsh border, Offa's Dyke itself being the best-known of all. They were all exceptional enterprises. That such efforts were put into defining frontiers, in a period when the emphasis of kingship was on overlordship of people rather than of territory, reveals the seriousness of the threats which produced them.

The other category of evidence from this phase of Celtic-Germanic confrontation is the occupation of defended hilltops, an important theme in settlement history throughout the Celtic areas of early Britain. From both specific excavated evidence and the implications of chance finds it has been estimated that at least 20% of Somerset hillforts were occupied in the early post-Roman period. No general figures are available for the other counties, though there are plenty of sites with interesting possibilities, such as Cadbury, near Tiverton in Devon, with its late Roman coin hoard.[9] Two of the best-known excavations on such sites were also in Somerset, at South Cadbury and Cadbury Congresbury. The massive refortification of the former in the fifth century is evidence of the powerful political authority of some British leader, and perhaps also of a shift of population from the more vulnerable urban centre at Ilchester. Cadbury Congresbury, the *burh* of St. Congar, reminds us also that some of these sites may have had religious uses. Some late Roman temples were situated in hill forts and some early Christian cemeteries are also found near them.[10] The motives for hilltop fortification in this period are probably wider than the purely military, but the threat of Saxon aggression and the general insecurity of these centuries were surely prime causes. The famous battle between British and Saxons at Mount Badon was said by Gildas to be a siege, that is it took place at a fortified site. It occurred around AD 500 somewhere in the west of England. At the same time, the nature of Celtic society makes it likely that internal conflicts had also a role to play. If the later history of Wales is any guide, the politics of the period is unlikely to have been simply a matter of Briton against Saxon. The tradition of occupying defended hilltop sites may have been due in part to the development of individual power-bases among the Celtic aristocracy. It was, after all, an aspect of Celtic lifestyle with its roots in prehistoric culture. Such internal conflicts are less known because they did not take recorded form as did those with the Saxons. But in a period when the protective framework of Roman army and government had disappeared, fear of strong neighbours, whatever their race, must have been ever-present. We do not know exactly who was responsible for these sites, whether their building was a practice controlled by the Dumnonian dynasty, or more widespread. It is likely that the use of such places was much more common than our available evidence reveals. Superficially hillforts occupied only in the pre-Roman Iron Age cannot be distinguished from those with post-Roman occupation. Moreover the excavations at South Cadbury revealed late Saxon and medieval phases of occupation, a reminder that the values of hillforts may have been apparent for many centuries. Another interesting example is Denbury, a hillfort in south Devon whose name means the *burh* of the men of Devon. Was this a major centre of resistance to the Saxons, or was it possibly of later date, occupied by the Saxons themselves?[11] In the Norman period some castles were built inside ancient hillforts, because they provided ready-made outer defences. It is interesting to speculate how far our lack of knowledge of early medieval rural settlements in the South West may be due to the defended sites having played a greater role than is evident. In the extreme west, in Cornwall, the tradition of the lightly defended 'rounds', some of which like Trethurgy near St. Austell, were occupied in this period, may bear this out.

Wessex

By the sixth century the West Saxons were well advanced into Dorset and Somerset, and in the seventh they took over Devon. Cornwall remained independent until the ninth and tenth centuries. By the ninth century, however, the West Saxons themselves were subject to serious inroads from Danish armies, attacking both overland and from the sea. In this context we have our first documented view of defended sites, a list of fortifications in and around Wessex known as the Burghal Hidage, reflecting the West Saxon defensive situation in Alfred's reign.[12] It included places of Roman origin (Exeter, Bath), sizeable places of Saxon origin with planned lay-outs (Wareham, Barnstaple) and other less planned sites (Lydford, Shaftesbury, Bredy, Christchurch, Lyng, Langport,Axbridge). Halwell (Devon), of unknown form, was succeeded in the tenth century by Totnes. Lydford was the most westerly of all, suggesting the west Saxons did not yet regard

Cornwall as an integrated part of their kingdom, or at least did not feel capable of defending it. Existing Roman defences provided obvious bases for resistance. Coastal locations, too, were important, reflecting the vulnerability of estuaries to Danish seaborne attack. The west country played a significant role in Alfred's resistance, Somerset being his refuge in the 870s, Athelney (together with Lyng a 'double-*burh*') the defended base from which he re-organized his efforts.

Care must be exercised when interpreting the significance of this list. It must not be assumed that the defensive functions of all places named were new. Wareham was probably already a place of assembly for armies. Some *burhs* were ancient Roman cities. Equally, there is a tendency to see it as an exhaustive list of towns, one of whose features was defence. But certain places, such as Dorchester and Ilchester, which beyond reasonable doubt had urban characteristics at this date, do not figure in the list. And it has recently been argued that there were many other, smaller urban places which did not figure.[13] The list is simply a record of the contemporary defences of Wessex. Some of these places came to have a broader urban character in the years that followed, but by no means all. Some did not develop economically and socially. Even where by the time of the Norman Conquest a flourishing town or city existed, this was often a relatively recent development. The documentary evidence of urban activity increases in the tenth century as does the evidence of the coins minted. The archaeological evidence of general urban occupation, as at Exeter, can be very late pre-Conquest indeed: it was not a flourishing town *circa* 900 despite its walls and its mint.[14]

Although these *burhs* could protect the rural population in time of attack as well as their permanent inhabitants, they were fundamentally the means by which the West Saxon dynasty protected its kingdom. Those who manned and maintained the walls, according to the evidence of the Burghal Hidage list itself, were raised in the surrounding countryside as a result of military obligation to the king, on the basis of the hidation of land. Indeed, such emphasis was placed on this feature of their organization that one might question whether the permanent population of these places *circa* 900 was of any significant size at all. As their other activities developed, the royal imprint continued to be fundamental: the coinage minted there was the king's, the justice dispensed there was the king's, and the portreeve (as well as the shire-reeve) was a royal officer. As the West Saxon kings reconquered the Danish areas in the tenth century, they established new *burhs*, or re-organized Danish settlements along their own lines. Admittedly, there was a strong physical contrast between the Saxon *burh* and the Norman castle, but their purposes were not always as totally different as is usually argued. Both could be used to protect a kingdom, as well to extend the power of the king into neighbouring territories. For Asser, Alfred's biographer, when the Danes entered it Wareham was a *castellum*, the normal later word for a Norman castle. The Normans and their successors often built defended towns in conjunction with castles to consolidate their conquests in Wales. Like the later castle, the *burh* was also a means of localizing defensive efforts. The king's army could not be everywhere, but locally raised garrisons, like the levies from the shires who met the Danes in the field, could defend crucial locations on the king's behalf.

The system seems to have worked well, despite the remark by Asser that Alfred had difficulty in getting people to work on defences.[15] Sometime between 878, when a Danish army occupied Exeter (and Wareham), and 893 when another army was unable to take it, Alfred had turned the crumbling Roman city into a defensible *burh*. It withstood a further attack in 1001, but was taken in 1003 because of the treachery of its reeve. In 997, a Danish army was active in the four south-western counties. Although it reached Tavistock and sacked the church there, it had to by-pass Lydford and did not take it. The ramparts, ditches, palisades and stone walls which have been excavated at Lydford, Wareham and South Cadbury, together with the surviving Roman fabric at Exeter (where, curiously, no late Saxon additions have ever been identified) give an impression of what the Danish armies faced.[16]

The last threat which these defences faced came from William the Conqueror in his south-western campaign of 1068, two years after his initial conquest of England.[17] Exeter's was the first of a series of challenges presented to the new king by several English towns, the repression of which from 1068-70 gave the Conquest so much of its violent and permanent character. The defences of Exeter certainly proved adequate to their task. The city held out for eighteen days, and was then not taken by force (despite the attempted undermining of the walls) but surrendered by its leading citizens. But what exactly motivated the rebellion? What did the men of Exeter see as the threat? The argument that the rebellion was part of a concerted plan to throw out the Conqueror and revive the old English kingdom cannot be sustained. There is no evidence for co-ordination with the other urban movements except that Exeter is said to have sent envoys to other cities. The resistance should be seen rather as regional and particularist. The citizens feared the new king would exact more from them than his predecessors. And despite the surrender Domesday Book reveals that Exeter's obligations indeed remained as before. The citizens seem to have made their point. Whether or not other west country *burhs* were also sites of resistance to the Conqueror is not known, though William certainly campaigned elsewhere in Devon and Cornwall after the fall of Exeter. The subsequent appearance of castles in Lydford, Totnes and possibly Barnstaple may suggest they too put up a fight. Initially most *burhs* remained in the king's hands, though Totnes was granted to a tenant-in-chief by 1086. Norman castles themselves certainly became objects of English attack. The men of Dorset and Somerset moved, unsuccessfully, against Montacute late in 1069, and there were probably other unrecorded incidents. But it was not simply an English versus Norman situation. This campaign was the first in which Englishmen served in William's army against fellow-Englishmen. A year later there was a further rebellion in which the men of Devon and Cornwall attacked Exeter itself, which remained loyal to the king. In fact the peninsula was not free of trouble for some time. Sons of the late king Harold, operating from the Viking ports in Ireland, attacked the Devon and Somerset coasts on several occasions, and in 1069 took a fleet to Exeter. By now any sense of 'national' confrontation was completely lost. The English province was being fought over by Irish Vikings led by the sons of an Anglo-Danish king on the one hand, and a Norman king, himself the descendant of another Viking dynasty, on the other. On top of which, William's chief representative in the repression of these continuing troubles was not even a Norman, but Brian, a count of Brittany.

The Norman Conquest

The aftermath of 1068 brought to the South West a new phenomenon, the castle. Allowing for the observations (above) on the similar purposes to which *burhs* and castles could sometimes be put, there were contrasts in other respects. Castles were smaller, they were residences of lords, their families and immediate followers, and were found not only in the king's hands but throughout the landholding class. They quickly became the symbol of the conquering aristocracy in the eyes of the English, and the construction work into which the latter were forced was detested, though it had plenty of precedent in the obligations owed to the Saxon kings. A major gap in our knowledge of medieval fortifications is the extent to which the residences of the late Saxon landowners were also defended. Two excavations in eastern England have revealed defended houses of this date beneath Norman castles, and the practice may have been more widespread than is realised.[18] The fact that both known sites were buried by Norman castles may be instructive: any amount of such evidence may lie hidden in the same way. Even if the living habits of the Saxons and Normans may have been less different in principle than commonly argued, there was certainly a major difference of degree. The scale, numbers and oppressive character of these sites struck deep in the hearts of the English, and whereas the homes of the Saxons, whatever their form, had been passed over in relative silence by the chronicles, the castle became a recurrent theme in post-Conquest sources. It dominated warfare, was a major element in political wranglings, and rapidly became the norm of aristocratic residence. Castles could be used in different contexts with different purposes, but they had a common ground in a society where violence, on many levels, was an ever-present threat.

It was in conquest and repression that the castle appeared first, as at Exeter in 1068, where William himself chose the site within the city walls. There followed closely from this twenty years in which the new king and his followers consolidated their hold on conquered lands.[19] It is often impossible to know exactly when a particular castle was established in this period, to distinguish an early castle of conquest from one arising a few years later in the redistribution of land. Exeter and Montacute are referred to in 1068-69, but otherwise the dating is more general. There are specific references in Domesday Book to Corfe, Dunster, Okehampton, Launceston and Trematon, and implications of Domesday or other documentary evidence which probably put Totnes, Lydford, Barnstaple, Stogursey and Wareham this early. A certain amount of land exchange was conducted in order to acquire suitable sites for castle building: Robert of Mortain at Launceston and Montacute, the king himself at Corfe (called, confusingly, by Domesday Book, 'the castle of Wareham'). There also appeared an unknown number of timber castles, whose surviving earthworks are commonly undateable within precise limits. These are very common all over England. Some were probably early, but their building continued for several generations. At Neroche, Somerset, the castle's central location in the neck of the south-west peninsula, in an area where Robert of Mortain had many estates, may suggest an early date and a role in the campaigns of conquest. Blackdown Rings, Loddiswell, south Devon, was built in the corner of an Iron Age hillfort. Domesday Book records the area to the west as one of lowered manorial values, a common sign of heavy fighting and the passage of armies. This too may have been a base for early campaigns. There are other examples where timber castles may have emerged as the defended residences and administrative centres of tenants-in-chief, such as Holwell, Parracombe, north Devon, in the lands of William of Falaise. These were the equivalent of the documented castles built by richer families elsewhere: Baldwin de Meules, sheriff of Devon, at Okehampton; Robert of Mortain at Launceston; Judhel, the Breton, at Totnes. There was no general pattern in this building activity, either in terms of location or of design. The individualistic designs remind us of the domestic aspect of the castle, as well as of the varied inputs into their conception: the gatehouse at Exeter contains Saxon architectural features as well as Norman. There was no overall strategy of distribution. For the most part castles were built as a result of many and particular circumstances. The relatively unusual grant of the Saxon *burh* at Totnes to Judhel, together with a landed endowment which almost approached the character of a castlery, presumably reflects a desire to protect the vulnerable Dart estuary. The nearest approach to any 'schemes' were the consistent building of castles in Saxon *burhs* all over England, and the delegation of castle-building powers to the lords of the Welsh borderlands.

The Anglo-Norman Period.

It was not only the Norman Conquest which saw the spread of castles. They continued to be built, for a variety of reasons, for several centuries, and were crucial to the processes of political and military life. The accession of Henry I in 1100 saw the arrival in England of men who had supported him in Normandy in the previous decade. One such man made his second appearance in the west country, the Breton, Judhel, who had been lord of Totnes in 1086. Evicted from Totnes by William Rufus, he was reinstated by Henry I, but now given the former royal town of Barnstaple and the lands of various former tenants-in-chief in north Devon. He died an old man in the 1120s, surely the only adventurer of the period to have held in succession two former Saxon *burhs*. The present castle at Barnstaple appeared in his time, though it is possible that it succeeded an earlier royal one.[20] The other notable arrival was Richard de Redvers, whose Norman family were given extensive lands in Devon as well as the Isle of Wight. In Devon they made the former royal manor of Plympton their *caput*, and established a castle and borough there. The endowment of these two men may reflect the new king's concern with the security of the west country, since both castles built were on major estuaries, and in the Isle of Wight Richard de Redvers inherited a castle (Carisbrooke) and castlery established in this sensitive area in the reign of the Conqueror.[21]

In the civil war of Stephen's reign, the castle had something of a hey-day. The main chronicle source, the *Gesta Stephani*, is full of accounts of sieges, of castles old and new, many of the latter held or built in defiance of the king, and is very informative on south-western events.[22] Some castles make their first documented appearance in these years: Bampton, Great Torrington (in Devon), Cary, Harptree, Taunton (in Somerset), Lulworth, Old Sherborne (in Dorset). Sieges are referred to at these and other places: Exeter, Plympton, Dunster and Corfe. Wareham retained a role of crucial importance to the Angevin cause in the west country, as the port which provided the link with France. Some castles were used as the bases for major rebellions against the king, as when Baldwin de Redvers held Exeter and Plympton against Stephen in 1136. Others were used as bases for the extension of local powers and influence, as when Robert of Bampton took the opportunity to redress long-held grievances over land and other matters.[23] Sometimes the individuals concerned disappeared quickly from view, as did Robert of Bampton and Alured (son of Judhel) of Barnstaple. Others achieved their ends: Baldwin de Redvers eventually regained control of Exeter and was created earl of Devon by Mathilda in 1141 (just as Reginald, a bastard son of Henry I, became earl of Cornwall, and William de Mohun became earl of Somerset and Dorset). Stephen's only major ally in the region was Henry de Tracy, whom he installed at Barnstaple after Alured's fall. He fought a lonely struggle from his north Devon lands against the king's many adversaries. A microcosm of the local confrontation may be seen at the mid-Devon village of Winkleigh. Here there were two manors, one held by tenants of the earl of Gloucester (together with Mathilda,Stephen's major opponent), the other held by the Tracy family of north Devon. The two earthwork castles which face each other represent either an unrecorded siege or simply the hostile lords of adjacent manors digging themselves in. The ownership of other manors in Devon containing undocumented earthworks may also point to similar castle-building activities.[24]

In the later twelfth and thirteenth centuries, royal interest in internal security is well reflected in expenditure on royal castles. In the last year of John's reign, for example, expenditure on garrisons and provisions, as well as the placing of castles in the hands of trusted men, reflects the king's concern with the growing opposition to him. Activity at Launceston, Exeter, Dunster and Corfe figures in this context.[25] In the 1220s events in the South West well reflect the difficulty experienced by Henry III's government in regaining control of castles which John had alienated. At the same time that Bedford castle was under siege in 1224, lesser operations were carried out at Plympton and Stogursey. All three castles were in the hands of Fawkes de Bréaute, one of John's captains who had received extensive rewards. Bedford was a royal castle. Plympton and Stogursey were in his hands because among his rewards had been the heiress to those two baronies. All three sieges were successful but involved considerable outlay. That at Plympton was conducted on the king's behalf by the sheriff of Devon. In the reduction of the defences of Barnstaple and Great Torrington castles in 1228 may be seen a final reflection of the troubles of John's reign. The work was ordered on the grounds that the castles were unlicensed, though in fact they had existed for a century. What was probably unlicensed was a more recent development of their defences.[26]

The Later Middle Ages.

There is a view which sees the later castles, despite their defences, as products of a nostalgic building tradition, of a desire to display wealth and ritual.[27] This view is easily exaggerated. What changed was the role of the castle in warfare and the emphasis within its design. It no longer dominated warfare. By *circa* 1300 castle-builders had developed their designs so well that the major siege warfare which had dominated earlier centuries was less common (though by no means absent). With the conquest of North Wales by Edward I the English kings reached a limit to their expansion and thereafter the history of royal castles was largely one of maintenance and domestic development. Nevertheless, the land-owning classes remained active in castle building, and they maintained traditionally defensive plans. But within those plans the emphasis was increasingly on domestic comfort and grand appearance. This was not without precedent. In the thirteenth century Richard, earl of Cornwall had developed Tintagel castle as a conscious evocation of the earlier traditions of Cornish kingship. It was built in tradional, defensive fashion, even though defence was not its purpose.[28]

In the early fourteenth century, Hugh II Courtenay rebuilt Okehampton, Tiverton and possibly parts of Plympton as part of the self-aggrandizement which led to his being made earl of Devon in 1335. Despite their defences, the emphasis at Okehampton and Tiverton was certainly domestic. At Plympton the work may have been designed to make the site look like Launceston, castle of the neighbouring earls and dukes of Cornwall.[29] In 1373, when John de la Mare received a licence to crenellate and built Nunney, he built his apartments in a tall, great tower, traditional in one sense but novel in the way it was put together and perhaps French in inspiration. In the same decade Sir Thomas de Hungerford built a very domestic castle at Farleigh Hungerford (more military features were added in the following century by his son Walter).[30] Compton, Devon, was the latest local example of domestic needs overpowering the defences. Here the fourteenth-century house was hardly defended, but was given various military trappings up to *circa* 1500.[31]

The violence which the builders of this period anticipated was not that of the great sieges of earlier days. Nevertheless, they felt a need for security, for protection of their families, followers and possessions. William Asthorpe came to Devon society and politics in the 1370s through marriage into the Dynham family. He became a member of parliament and sheriff of Devon, but the period witnessed much friction between himself and local society, including violent affrays with men of the abbot of Dunkswell and various pieces of litigation. The licence to crenellate which he acquired in 1380 for his property at Hemyock probably had a dual function: to demonstrate his social advancement in the county and perhaps also to develop a secure home against the local enemies he was making. His little castle looked quite impressive, but the surviving fabric suggests the defences were not very practicable. A real example of domestic violence occurred somewhat later in south Devon. In 1428, on the death of Thomas Pomeroy, his cousin Edward and his family were violently expelled from their house at Berry Pomeroy by a rival family faction who broke into the enclosed courtyard. It was not until later in the century, however, that Berry was turned into a more defensible castle (see below).[32]

In 1455 the Courtenays launched an attack from Tiverton on the undefended property at Upcott (near Cheriton Fitzpaine) belonging to Nicholas Radford, lawyer of their rivals the Bonvilles.[33] Radford's murder scandalized contemporary society and led to criminal proceedings. It may be cited as an extreme example of the threats against which men of this period felt in need of defence. Devon and Cornwall have been called the most lawless areas outside the far north of England during the Wars of the Roses, too easily thought of in dynastic terms but in which the escalation of local animosity was a more pressing problem for much of society. In February 1427 William Bonville of Shute took an armed party and broke into Thomas Brooke's property near Axminster, assaulted his servants and caused general damage. The following May, Brooke sought a licence to crenellate for his house at Wycroft near Axminster, and this was presumably not a coincidence. It reflected his desire for security, and also his desire for favour at the royal court. In the latter he seems to have been successful, since associated with him in the licence were a duke, two earls and five other knights.

The recent tendency to write off licences to crenellate as mere expressions of a pompous and chivalric society has gone a little too far.[34] Certainly they were prestigious, they served the recipients' interests, and many licences were granted in circumstances devoid of military or political content. Many recipients of licences in Devon, for example, had been loyal servants of the king in the locality: sheriffs, royal butlers in the ports, controllers of customs, escheators, as well as members of parliament. But some licences may have reflected local tensions, and others may have reflected fears of external attack (see below). Certainly licences to crenellate cannot be used as a general guide to castle-building, since many building operations seem to have occurred without their receipt. As an example of the latter may be cited Powderham, probably developed from the 1390's by

Philip Courtenay (of the junior branch of the Courtenays of Okehampton). The medieval fabric is now buried within the later house, and there are no obvious signs of serious defences. Nevertheless it was no 'toy castle' for in November-December 1455 it held out against a siege of nearly two months, withstanding attacks from archers, crossbowmen and gunners, and was not taken. In the account of the episode it was not even called a castle, but simply a house. The Courtenay-Bonnville feud, of which Radford's murder and the attack on Powderham were by-products, involved castles and other houses in various ways. The earls of Devon were lords of Okehampton, Plympton and Tiverton, their junior branch (allies of the Bonvilles) occupants of Powderham. The Bonvilles were stewards of the Duchy of Cornwall (and therefore keepers of Lydford castle), as well as constables of Exeter and Taunton castles. In 1451 there was a brief attack on Taunton castle, and the mid-1450s saw the Courtenay occupation of Exeter city and their attack on Powderham. The Bonvilles responded with an attack on the Courtenay house at Colcombe, and the final field engagement at Clyst St. Mary was followed by the ransacking of the Bonville house at Shute (itself partly fortified) by the victorious Courtenays. The late medieval castle may indeed have been a place of chivalric pageant, lavish entertainment and ceremonial. But it was also a place which could sometimes witness real violence.

The threat from overseas.

Powderham lies on the Exe estuary, and during its siege the Bonville party made an unsuccessful counter-attack across the river from Lympstone. Coastal violence and problems of coastal security were naturally vital to a region with such a long coastline. These problems are reflected in the location of many earlier sites, Roman fortifications, Saxon *burhs*, Norman castles. It was not only defence of sensitive points of entry which mattered, but also communication around the coasts. In 1221 Fawkes de Bréaute attempted to provision his castle at Plympton by sea, but the boats carrying grain from Exminster were intercepted by the sheriff of Devon.[35] In the same period, the north Devon and Somerset coasts had proved a problem to which king John had addressed himself in person. His efforts to grant the island of Lundy to the Knights Templars in defiance of its occupants, the family of William Marsh, led to measures for the defence of the coast and attacks on Lundy itself. Although Marsh returned to royal favour in 1204, his more famous outlaw descendant, another William, used the island as his base from 1235-42. It was in the context of the earlier episode that John granted permission to Alan of Hartland to fortify his house and defend it by force against William Marsh. It is not only one of the earliest surviving records of royal permission for castle building in the region, but it is also an interesting reflection of a general theme: coastal defence through fortification carried out by individuals but with some element of royal intervention.[36] This presumably underlay the foundation of some earlier estuarine castles, and it recurs in later contexts of coastal defence.

It is largely in connexion with the threat of French raids and invasion that coastal defence is discussed in the later middle ages. This itself, of course, was not a new problem. In the later twelfth and early thirteenth centuries it was the protection of the kingdom from the French which led to such massive expenditure and development at Dover castle, 'the key of England' as the contemporary chronicler Matthew Paris called it.[37] It was also why other coastal areas were so sensitive, a fact reflected in the appearance of such early castles as Colchester in the east, or Corfe in the south. The so-called 'private' nature of royal castles should not disguise the important role they played in general defence. It has been pointed out that Queenborough, Kent, built by Edward III in the 1360s and commonly regarded as the last new royal castle of traditional character, was also the first of the specialized coastal fortifications. It included provision for artillery, and its location on the Isle of Sheppey was specifically intended to protect the Thames and Medway.[38] Equally, potential threats from abroad stimulated not only royal efforts but efforts by other individuals encouraged by kings, as at Rochester in the south east, or Carisbrooke in the south. South-western castles such as Totnes, Plympton and Trematon enjoyed at least some of this character. In the later middle ages we also find connexions between the building efforts of local society and external problems. William Asthorpe (above) was involved in the same year (1380) as he received his licence to crenellate for Hemyock, in a commission of array 'in readiness to resist foreign invasion'. Richard Merton received licences to crenellate for Great Torrington in 1340 and 1347 'in consideration of his good service in the war with France',[39] and the construction of Nunney in Somerset in 1373 (above) is said to have been financed by profits from the French wars. It was at the same time that Dartmouth was receiving its famous defences, from the 1480s, that the Pomeroy house at Berry near Totnes was transformed into a castle, incorporating gunports and using some of the same stone and other details as were employed at Dartmouth itself.[40] In 1403 a licence to crenellate was granted for a house on the opposite bank from Dartmouth at Gomerock or Kingswear.[41]

The second reflection of English reaction to outside threats was the development of town defences from the thirteenth century. The extent to which the major *burhs* maintained their defences after the Conquest is not known. Militarily, the landscape was dominated by castles, though occasionally towns were besieged, as were Bath and Bridport in the civil war in Stephen's reign. When southern towns became sensitive to French invasion in John's reign, the building of walls was achieved by financial intervention from the king.[42] At first this was piecemeal, with allowances of money or building materials, but by 1220 the murage grant had emerged, permission for the townsfolk to levy tolls on goods arriving for purposes of sale, for a specified number of years. The money thus raised supported building works organized by the urban oligarchies themselves. The recorded grants of murage are not a foolproof guide to the history of town walls. Some building operations were achieved without them, and some grants did not result in actual work carried out. Nevertheless we are very dependent upon them. We have written evidence for works at Exeter and Bristol from the mid-thirteenth century; at Totnes and Bridgewater in the later thirteenth century; at Melcombe (Dorset) in the early fourteenth century; at Plymouth, Bath and Wells in the later fourteenth century; and at Poole and Ilfracombe in the fifteenth century. However, there is no real evidence that anything was done at Ilfracombe, Melcombe and Wells, whereas Launceston, Barnstaple and Taunton certainly did have town defences though no known grants. It would be a mistake, however, to see the building of town walls simply as a response to fears of the French. Equally important was the symbolic role of defences, a reflection of the wealth and pride of the urban community. For this reason gates were not only strong but also visually impressive. In the South West, town defences were not generally put to the test in this period. Plymouth and Dartmouth were attacked and damaged by the French in 1403-1404, which may have encouraged the development of their fortifications (see below). Perhaps the best-known threat was an internal one: in 1549 the men of Devon and Cornwall who led the Prayer Book Rebellion laid siege to Exeter for five weeks.[43]

The third and most obvious reflection of outside problems was the beginning of what is usually called 'coastal defence'. This flowered in the sixteenth century, but the famous developments of Henry VIII's reign are beyond the scope of this essay.[44] The South West saw important developments in the fourteenth and fifteenth centuries which anticipated later ideas. These developments took place at Plymouth and Dartmouth.[45] The strategic value of the Tamar and Dart estuaries had long been appreciated. Dartmouth had seen the assembly of crusading fleets in 1147 and 1190, and was a collecting point for royal ships from 1204, being the embarkation point in 1206 for king John's expeditions to France. The Saxon *burh* and Norman castle at Totnes reflected the importance of the Dart estuary, as did the castles of Trematon and Plympton either side of the Tamar. In

1295 a royal expedition to Gascony set off from Plymouth, and in 1357 the Black Prince stayed at Plympton en route to the same area. Developments at both places in the later middle ages look both backwards to the traditions of earlier times, yet forwards to the sixteenth century, blurring the distinction between medieval and modern.

At Plymouth a murage grant for six years was received in 1377 for the purpose of fortifying the town.[46] In 1404, subsequent to a damaging French raid, Henry IV commissioned the prior of Plympton and the abbot of Tavistock to employ masons and carpenters to fortify the town with a wall and towers, though there seems to have been no direct input of royal finance.[47] Subsequently, there were further local initiatives, with various bishops of Exeter granting indulgences to those contributing to the work (Stafford in 1416, Lacy in 1449, Veysey in 1520).[48] So far the situation was rather medieval in character. But what in fact was built was not a town wall at all. It was, as the wording of bishop Lacy's indulgence stated, a castle (*castrum*) within the town (*villa*). It is shown on a map of *circa* 1540,[49] much as Leland described it, 'a strange castle quadrate, having at each corner a great round tower'.[50] Only a short stretch of wall is shown extending towards the town, and nothing suggests it was ever continued. One fragment of the fabric survives in modern Plymouth, much having disappeared in the nineteenth century. But neither was this fortification a traditional medieval castle. It was not the residence of a great lord, a variety of interests had been involved in its building, and the arrangements for its defence owed more to the urban than the castle background.[51] The mayor was in command, and the aldermen supervised the defence of the four main towers, whose maintenance was allocated to the four wards of the town. With its mixed character, stemming from the traditions of royal intervention and local initiative, defended by a community but not enclosing it, and on a coastal location, it represents a curious compromise between the castle, the defended town and the later coastal fort.

At Dartmouth in 1336, Edward III had commissioned Hugh Courtenay, earl of Devon, to see to the defence of the town against French invasion.[52] In 1374, 1377, 1381 and again in 1406, the mayor, John Hawley, was commissioned to arrange defences, and by 1388 'a fort by the sea' was under construction.[53] The resources for this work came from the locality, with the impetus, as in the early work at Plymouth, from the king. As also at Plymouth, it was neither a traditional castle nor a town wall which was constructed. It was a walled and towered enclosure, shown on the same sixteenth century map (and again on an eighteenth century one)[54], situated where the present, more famous fortification stands, and surviving beside it in fragmentary form. The 'old castle' was later referred to in relation to its successor, whose construction can be followed in the late fifteenth century.[55] In 1462 Edward IV granted moneys for twenty years for work on the defences.[56] This grant was later extended in 1481 and 1484, and again in 1486. The expenditure in the 1460s was specifically for the boats and chains for blocking the estuary, cannon and wages for troops. In 1481, work on 'a strong tower' for artillery and a chain across the river had begun and the rest of the expenditure can safely be related to the building known as Dartmouth Castle.[57] In its general form, a tall double tower, it looked back to the castle tradition, and parts of its plan were influenced by the earlier structure. But its sophisticated gunports were novel, with their large rectangular openings, and the overall design, intended exclusively for artillery and professional gunners, took it away from the medieval world of the fortified home. And while the work was carried out under local organization, the direct financial involvement of the crown (the initial money came from customs and subsidies from the ports of Exeter and Dartmouth, which would otherwise have been royal income) was a feature pointing the way forward. Dartmouth was easily assimilated, both in organization and function, into the system of coastal fortifications which Henry VIII developed around the eastern and southern coasts.

Conclusion

It is easy to become overwhelmed by the problems of outside danger and general defence, though they were certainly important at the beginning and end of the long period under review. For much of the time the day to day problems of security were posed as much from inside society as from outside it. Internal tension and weakness, local rivalries, failure of feudal loyalties, the threat of civil war: these formed the background to the appearance of hundreds of fortified residences in medieval England. There were at least twenty castles in each of the four south-western counties.[58] From minor defensible homes to great strongholds, they represented the normal aspirations of the wealthy landed class, became so much a habit that it is not necessary to seek specific military motives in the foundation of every one. By the seventeenth century the lawyer, Sir Edward Coke (1552-1634), could quote 'The house of everyone is to him as his castle and fortress', now known in the more familiar phrase 'An Englishman's home is his castle'.[59] It is a notion which has survived in our attitudes to privacy. And in the sixteenth and seventeenth centuries, when the building of castles had declined, the last vestiges of the medieval tradition were still to be seen in the gatehouses which commonly accompanied the houses of the gentry.[60]

NOTES

1. For the background to theories of kingship and government see W. Ullman's *A History of Political Thought: the Middle Ages* (1965) and *Principles of Government and Politics in the Middle Ages* (1961).
2. See C. Coulson, 'Rendability and castellation in medieval France', *Château Gaillard, Études de Castellologie médiévale*, VI (Caen 1973), 59-67, and literature referred to.
3. The literature goes back to the beginning of the century, especially E.J.Armitage, *Early Norman Castles of the British Isles* (1912); more recently R. Allen Brown, *English Castles* (3rd edn. 1976).
4. See M. Todd, *The South West to AD 1000* (1987), ch.10.
5. M. Todd (personal communication).
6. C. Taylor, *Dorset* (1970), 42-44.
7. C. and A. Fox, 'Wansdyke reconsidered', *Arch. J.* 115 (1958), 1-48.
8. W.G.V. Balchin, *The Cornish Landscape* (1983 edn.), 66.
9. P.J.Fowler, 'Hillforts, AD 400-700', in *The Iron Age and its hillforts* (ed. M. Jesson, D.Hill, 1971), 203-213.
10. M. Aston, I. Burrow (eds.), *The Archaeology of Somerset* (1982), 83-107; I. Burrow, *Hillfort and Hill-Top settlement in Somerset in the first millennium AD* (B.A.R., 91, 1981).
11. For Devon, see the useful map by I. Burrow in *C.B.A. Arch.Rev.* (*Gps xii,xiii*), no. 7 (1972), 46-47.
12. R.H.C. Davis, 'Alfred and Guthrum's frontier', *EHR*, 97 (1982), 803-810; D.Hill, 'The burghal hidage: establishment of a text', *Med. Arch.* 13 (1969), 84-92.
13. For Wareham, see D. Hinton, R. Hodges, 'Excavations in Wareham, 1974-75', *Proc. Dorset Nat. Hist. Soc.* 99 (1977), 42-83; see also the various case-studies in J. Haslam (ed.) *Anglo-Saxon Towns in Southern England* (1984).
14. J.P.Allan, C.G.Henderson, R.A.Higham, 'Saxon Exeter', *ibid.*, 385-414.
15. W.H. Stevenson (ed.) *Asser's Life of King Alfred* (1959), 36 (Wareham), 77 (general defences). The following incidents are all described in *The Anglo-Saxon Chronicle*, ed. D. Whitelock (1961).
16. For the various results of excavations, see references in Haslam (ed.) 1984 (note 13).
17. The fullest narrative of what follows is in Orderic Vitalis, *The Ecclesiastical History*, ed. M. Chibnall (Oxford Medieval Texts) vol. II (1969), 210-224.
18. G. Beresford, 'Goltho Manor, Lincolnshire: the buildings and their surrounding defences, 850-1150', *Proc. Battle Conf. IV* (1981), 13-36; B.K. Davison, 'Excavations at Sulgrave, Northamptonshire, 1960-76: an interim report', *Arch. J.* 134 (1977), 105-114.
19. For what follows see D. Renn, *Norman Castles in Britain* (1971); R.A.Higham, The Castles of Medieval Devon (unpub. Ph.D. thesis, Exeter 1979); R.A.Higham, 'Early Castles in Devon (1068-1201)', *Château Gaillard, Études de Castellologie*

médiévale, IX-X (1982), 101-116; R.A.Higham, 'Castles in Devon', in *Archaeology of the Devon Landscape*, ed. S.C.Timms (Devon County Council, 1980), 70-80; B.K. Davison, 'Castle Neroche: an abandoned Norman Fortress in South Somerset', *Proc. Somerset Arch. Nat. Hist. Soc.*, 116 (1972), 16-58; A.D.Saunders, interim reports on Launceston, in *Cornish Arch.*, 9 (1970), 83-92; 16 (1977), 129-137.

20. See R.A.Higham, 'The origins of Barnstaple castle', *Proc. Devon Arch. Soc.*, forthcoming.

21. R.A.Higham, S. Goddard, M. Rouillard, 'Plympton Castle, Devon', *Proc. Devon Arch. Soc.*, 43 (1985), 59-75.

22. K.R.Potter, R.H.C. Davis (eds.), *Gesta Stephani* (Oxford Med. Texts, 1976).

23, See A. Hamlin and R.A.Higham, 'Bampton Castle', *Proc.Devon Arch.Soc.* forthcoming.

24. See Higham 1982 (note 19).

25. R. Allen Brown, H.M. Colvin, A.J.Taylor, *The History of the King's Works*, vol. II (1963), passim; further information can be found in the various volumes of chancery enrolments for John's reign, published by the Record Commission.

26. For the general background, see F.M.Powicke, *King Henry III and the Lord Edward* (2 vols. 1947); for Plympton, see note 21; for Barnstaple and Great Torrington, see *Close Rolls, 1227-31*, 69-70.

27. For example, C. Platt, *The Castle in Medieval England England and Wales* (1982), chapters 7 and 8; R. Allen Brown, *English castles* (1976), chapter 7.

28. O. Padel, 'Tintagel: an alternative view', Appendix 2 in C. Thomas, *A Provisional list of Imported Pottery in Post-Roman Western Britain and Ireland* (1981), 28-29.

29. For Plympton, see note 21; for Okehampton and Tiverton, see R.A.Higham, J.P. Allan, S.R.Blaylock, 'Excavations at Okehampton castle, Devon; Part 2:the bailey', *Proc. Devon Arch. Soc.*, 40 (1982), 19-151; R.A.Higham, *Okehampton Castle: Official Handbook* (H.M.S.0. 1984).

30. R. Allen Brown, *English Castles* (1976), 137; C. Platt, *The Castle in Medieval England and Wales* (1982), 121-125.

31. A. W. Everett, 'Compton Castle', *Trans. Dev. Ass.*, lxxi (1939), 343-345, and 'The rebuilding of the hall at Compton Castle', *ibid*, lxxxviii (1956), 75-85.

32. See Higham 1979 (note 19), 155, 263 (Hemyock), 87 (Berry Pomeroy); *Cal. Pat. Rolls, 1377-81*, 552; PRO, KB9/223/2, mem.38.

33. For what follows, see R.L. Storey, *The End of the House of Lancaster* (1966),84-87, 165; for Wycroft, Higham 1979 (note 19), 263-264, *Cal. Pat. Rolls, 1422-29*, 400; for Powderham, PRO, KB9/16/mem.65; for a list of licences to crenellate in Devon, Higham, *ibid.*, 50-51, and R.A. Higham, 'Devon castles: an annotated list', *Proc. Devon Arch. Soc.*, forthcoming. I am very grateful to Dr. M. Cherry and Dr. C. Tyldesley for discussion of Devon's late medieval families.

34. for example, C.Coulson, 'Structural Symbolism in medieval castle architecture', *J.B.A.A.* 132 (1979), 73-90.

35. See note 21.

36. Higham 1979 (note 19), 148-149.

37. *King's Works*, II, 629.

38. J. R. Kenyon, 'Early Artillery Fortifications in England and Wales: a preliminary survey and re-appraisal', *Arch. J.* 138 (1981), 205-240, esp. 209.

39. *Cal. Pat. Rolls 1345-48*, 228.

40. Higham 1979 (note 19), 86-88; Kenyon 1981 (note 38), 227.

41. *Cal. Pat. Rolls, 1401-05*, 219.

42. See generally H.L. Turner, *Town Defences in England and Wales* (1971), esp. Part V; S. Reynolds, *An Introduction to the History of English Medieval Towns* (1977); also T.P. Smith, 'Why did medieval towns have walls?' *Curr. Arch.*, 95 (1985), 376-79.

43. J.A.Youings, 'The South-Western Rebellion of 1549', in *Southern History*, 1 (1979), 99-122.

44. See *The History of the King's Works, vol. IV, 1485-1660*, part II, ed. H.M. Colvin (1982); B.H. St. J. O'Neil, *Castles and Cannon* (1960); A.D. Saunders, 'The Coastal Defences of Cornwall', *Arch. J.* 130 (1973), 232-237; Kenyon 1981 (note 38).

45. For the background, see Higham 1979 (note 19), 267, 172-176; for Dartmouth, B.H. St. J. O'Neil, 'Dartmouth Castle and other defences of Dartmouth Haven', *Archaeologia*, 85 (1935), 129-157; R.N.Worth, *History of Plymouth* (1890), 402-423.

46. *Cal. Pat. Rolls, 1377-81*, 81.

47. *ibid., 1401-05*, 353.

48. G. Oliver (ed.), *Monasticon Diocesis Exoniensis* (1846), 131.

49. Brit. Lib. Cott. Aug. I,i,39.

50. L. Toulmin-Smith (ed.), *The Itinerary of John Leland, vol. I* (1907),214.

51. R.N. Worth, 'The ancient castle of Plymouth', *JBAA*, 39 (1883), 255-258.

52. T. Rymer, *Feodera* (ed. Caley, Holbrooke, Clarke) II (1821), pt. ii, 951.

53. *Cal.Pat. Rolls, 1374-77*, 32, 486; *Cal. Close Rolls, 1385-89*, 537-8.

54. See note 49; Brit. Lib. King's MS 45, fol.51; O'Neil 1935 (see note 45)

55. *Cal. Pat. Rolls, 1476-85*, 251; Leland (see note 50), 221; *Letters and Papers, Foreign and Domestic of the reign of Henry VIII*, vol. 3, pt. 3, 997.

56. *Cal. Pat. Rolls, 1461-7*, 75.

57. *Cal. Pat. Rolls, 1476-85*, 251.

58. For Devon, see Higham 1979 (note 19) and Higham, forthcoming (note 33); I am grateful for figures provided for Cornwall (N. Johnson), Somerset (R. Croft) and Dorset (L. Keen).

59. *Oxford Dictionary of Quotations* (3rd edn. 1979), 154.

60. See various contributions by the author to the revised edition of Pevsner's *Buildings of England: Devon* (ed. B. Cherry, forthcoming).

Bowmen, Billmen and Hackbutters: the Elizabethan Militia in the South West (The Harte Lecture 1986)

BY JOYCE YOUINGS

On 20 May 1580 at Bradninch, not far from Exeter, the able-bodied men of the parish were assembled and their names recorded on a skin of parchment. Those of the town were listed separately from those of 'the manor' and all were further categorized according to their supposed aptitude with weapons. First came the harquebusiers, or in common parlance 'hackbutters',* scheduled to use the relatively new match-lock hand guns, five of them in the town and only one in the rural area. Next came the bowmen, the seven in the town including John Holwell, master tanner, two weavers and a smith, with thirteen of these necessarily muscular men among the countrymen. Those who followed the plough were traditionally thought to make the best archers. Forty in all were classified as pikemen, presumably all tall fellows capable of managing weapons which were 16-18 feet in length, and 63 as billmen, smaller chaps who would fight with the shorter shafts, tipped with hooks. Finally there were 36 unarmed 'pioneers', men handy with picks and shovels, bringing Bradninch's total able manpower to 165, out of a total population probably of well over one thousand, not counting vagrants and the very poor.[1]

Only for the year 1569 do there survive similar lists of names, but for that year we have fairly complete coverage, parish by parish, hundred by

* Technically the hackbut was an earlier version of the harquebus. Both were about 3 feet long and weighed about 10 lbs, being fired by igniting gunpowder by means of lighted match. Held close to the cheek, in unskilled hands they were a greater danger to the bearer than to his target.

hundred, for each of the four south-western counties.[2] The Bradninch roll of 1580 makes no mention of any resident gentlemen or other substantial inhabitants but in 1569 Peter Sainthill gentleman, though not included among the able-bodied, was assessed by virtue of his notional landed income to provide one light horse, one corslet (the usual armour for a pikeman), one pike, two suits of 'almain rivets' (flexible body armour, presumably of German manufacture and now somewhat outdated), one hackbutt, two bows with arrows and two steel caps. Thirty-nine other Bradninch people, including Joan Mortimer and three other women, presumably well-to-do widows, were assessed for smaller amounts of arms and armour, either on their income from land or alternatively on the value of their movable goods. About half the men so 'presented' to make contributions in kind were also listed as able, but in all but one case as bowmen or billmen. Presumably the better-off parishioners managed to avoid risking injuring themselves fiddling with dangerous firearms. There was also a parish stock of arms and armour which by 1580 included two calivers, the rather newer and longer handguns, but also a great many more bows and arrows than in 1569.[3] Assuming that all the weapons were serviceable there were just about enough to go round, except for the billmen, but they could make do with the almost identical farm implements which they used for trimming hedges.

Unlike the rogues and vagabonds who were conscripted in the counties for continental campaigns and for service in Ireland, the militia, at least in practice, was a hand-picked army. In July 1577 the commissioners of musters for Devon were instructed to include only 'meet and able husbandmen and farmers' sons that are likely to continue in the place' and not 'such artificers as commonly are removing', in other words established householders and their adult sons and servants, not the vagrants or mobile, masterless workmen of whom Tudor politicians had such obsessive suspicion. Moreover only those assessed for parliamentary taxation at £5 or more were required to supply arms and armour, but those same instructions of 1577 extended this particular net to include

> such persons as by keeping of taverns and alehouses do gain highly by resort of persons more for pleasure than necessity.

Indeed in 1570 the earl of Bedford, as Lord Lieutenant, was specifically instructed not to be satisfied with people's rating in the subsidy rolls but to take due note of their 'open doings in the world' and to endeavour to discover those 'of secret wealth and never charged with service as the gentlemen be', especially in regard to the provision of horses.[4] Gentlemen were, by definition, those who not only lived like gentlemen but did so openly.

This last injunction was no doubt intended to underline a clause in

Queen Mary's Militia Act of 1558, the statute which formed the basis of the Elizabethan system. This repeated a statutory provision of Henry VIII's reign, that any man whose wife wore

> any gown of silk ... any French hood or bonnet of velvet ... or any chain of gold about her neck or in her partlet ... or any petticoat of silk ...

should provide an additional 'stoned' light horse, what was called in more polite circles a gelding. The obligation ceased in the event of divorce. There were remarkably few in Devon who in 1569 were surcharged for their 'wives apparell'. John Parker esquire of North Molton, was assessed, *inter alia*, at a light horse, but not on account of his wife's apparell. He was, in fact, the richest landowner in the parish, his tally of arms and armour surpassing that of Peter Sainthill over at Bradninch by just one morion (a visorless helmet).[5] One cannot, I think, conclude either that the ladies of Devon were all as homely in their attire as they are made to appear in the muster rolls, or that more discreet inquiries on the part of the muster commissioners would have solved the seemingly perpetual problem of a shortage of horses.

As a Justice of the Peace, John Parker was exempt from being 'abled'. The bishop, as a lord spiritual, was not even required to supply arms and armour, but the rest of the clergy were called upon to make their contribution. An elaborate assessment of the whole diocese made in 1588 shows the Dean of Exeter set down to provide one horse, fully 'furnished', two corslets and a caliver, and each of the archdeacons a horse and musket. Mr Phillips, rector of North Lew, was assessed at half a horse and the vicars of Okehampton and Dean Prior the other half between them. All were, in fact, commuted into cash.[6]

In a highly-centralized and personalized monarchical regime such as prevailed in Tudor England ultimate responsibility for protecting the realm from the danger of civil disturbance or the horror of foreign invasion lay at Westminster and whatever action was called for was by the authority of the Lords of the Privy Council and, indirectly, of the sovereign. Moreover Queen Elizabeth and her ministers insisted on being kept very well informed, for which, in the almost total absence of local administrative records, historians must be grateful. The ancient tradition, merely confirmed as long ago as 1285 in the Statute of Winchester, that all able men between the ages of 16 and 60 must be ready to defend their home territory against both internal and external enemies could only be made effective with the co-operation of the Crown's local agents, the knights and leading gentlemen. The Privy Council could encourage and prod and at times express its extreme dissatisfaction, but it had to be careful not to exhaust either the pockets or the patience of the people who mattered locally. It also had to defer to local knowledge. Nowhere is this better seen than in the two great changes which came over the militia in the latter half of the sixteenth century, the selection from those 'abled' of companies or bands thought sufficient and affordable for current needs and the introduction of regular military training.[7]

The Queen's accession in 1558 was quickly followed by an order for a general muster. The eleven named Devon JPs who gathered up the details excused the late return of their brief certificate by the fact that some had been attending the Queen's coronation and that others were aged, sick and unable to travel far from home. However on 9 March 1559 they certified a total of 6,762, an average of barely 15 per parish, of whom only 44 were gunners, 2,482 being archers and 4,236 billmen. Among the weapons listed were 136 hackbuts, 2,426 longbows and 2,175 bills. Their opposite numbers in Cornwall had pointed out in February that most of the able men in the coastal parishes were mariners and fishermen, whose services in the event of an emergency would be claimed by the Lord Admiral, and that the inland dwellers were mostly tinners who traditionally mustered separately under the Lord Warden of the Stannaries. They reported a total muster of only 930 able men. Dorset and Somerset each returned totals of over four thousand, but in 1560 Dorset sent a revised total of over nine thousand, of which half were described as 'unable' but 'willing'. By then Devon had more than quadrupled its total to over 28,000 but divided into a principal force of 5,284, a reserve of 5,244 and a residue of nearly 17,500 described as 'unable but to keep the country'.[8] The militia was not, of course, expected to cross the county boundary except in the event of an actual invasion, and never in order to quell a merely domestic uprising.

In the unusually detailed returns of 1569 Devon returned a considerably smaller total of under 15,000 able men, including about 450 separately abled in the city of Exeter, but excluding the town of Plymouth and the tinners. Cornwall returned some 12,000, including its tinners. In Devon the categories of billmen, bowmen, pikemen and hackbutters were approximately of equal size, the number of shot having been considerably increased since 1559-60, but in Cornwall billmen still accounted for about half the total and there were still less than 500 hackbutters. Some of those listed in the hundreds of Penwith and Kerrier were armed only with slings. Given the chance they might even have confused an enemy task force into thinking that the descendants of the victors of Crecy had produced another secret weapon. However, throughout the South West in 1569, except for a few missing parishes, the men of ability, both in substance and physique, for one brief moment in early modern history stand revealed to us. But anyone contemplating a computer-assisted analysis of population density should note that the able men of Sherwill in north Devon included a fair sprinkling of the same names as those listed for the adjoining parish of Lynton.[9]

Responsibility for the musters was usually committed to a select group of local magistrates. In 1569 there were six for Devon, Sir John St Leger from Monkleigh in north Devon (the landowner, as it happens, with the largest assessment in the county for arms and armour), Sir Arthur Champernon of Dartington (one of the busiest of the young Queen's local agents), Sir John Moore of Cullompton, Sir Peter Edgcombe of Cotehele (actually resident in Cornwall but in this as on many other occasions called in to supply the want of men of standing in south west Devon), Sir Gawen Carew of Mohuns Ottery in east Devon and William Strode esquire of Plympton St Mary, the current sheriff. Attendance at musters being a statutory duty, responsibility for dealing with defaulters was a matter for Quarter Sessions, the penalty being a fine of 40s or ten days' imprisonment. Only from 1592 when judicial records begin for Devon is it possible to identify offenders and they turn out to be remarkably few. In 1594 James Meare and Nicholas Hill were gaoled for 'running away ... being warned for the muster', but any other miscreants were presumably dealt with by their parish constables. In 1598 a certain John Bridgman, who had lost a hand while in training at Cullompton, was awarded a pension of £5 a year from the county treasury.[10]

In times of crisis the JPs had looming over them the far greater power and prestige, especially in matters relating to the county's defences, of the Lord Lieutenant and his deputies. The development of the office owed a very great deal to the responsibility for the South West laid by Henry VIII and Edward VI, at a time when it was regarded as the most refractory part of the whole of England, on John Russell, first earl of Bedford, and subsequently, after a short interval early in Mary's reign, by Mary herself and by Queen Elizabeth on his son Francis, the second earl. Francis, although a man later noted for his Puritan sympathies, was actually appointed as Lord Lieutenant (of Devon, the city of Exeter, Cornwall and Dorset) by Queen Mary early in 1558. Elizabeth renewed his commission in May 1559 but only for Devon, Cornwall and the city of Exeter. She was very loathe to make the office permanent, let alone hereditary, and in 1561 announced firmly that, the country being peaceful, there would be no commissions of lieutenancy for the time being. But in 1570, in the wake of the Northern Rebellion, Bedford was back in office. More important, perhaps, he now had deputies, Sir John St Leger and Sir Gawen Carew for Devon and Hugh Trevanyon esquire for Cornwall. Although, as far as I know, never actually commissioned as such, the mayor of Exeter acted as the Lord Lieutenant's deputy within the bounds of the county borough. Bedford's immediate task in 1570 was to put the two counties and the city into a state of readiness to repel any attempt at invasion by the French. By the end of the year he was being thanked and stood down, but by 1574 he was back in office, with similar instructions except that the enemy was now Spain. As never before the South West was now in the front line and, except for a short interval in 1585-6, the office of Lord Lieutenant was thereafter continuously filled.[11]

Bedford had many other commitments, including attendance at meetings of the Privy Council, but when his presence was required he had convenient residences in Exeter and Tavistock, both part of the large grant of former monastic property made by Henry VIII to his father in 1539. The first earl, as President of the short-lived Council of the West, had been given the complimentary duty of pacifying the region by acting as arbiter in private disputes and it is interesting to note that in the instructions issued to the second earl in 1570 there is a somewhat veiled reference to the exercise of 'mediation ... wherein surely much good may be procured and occasion given to concord and quiet'. He had a special responsibility, along with the Assize Judges, of overseeing the magistrates in the exercise of their multifarious duties. In times of actual hostilities he was also authorised to exercise martial law, and this enabled him, in the words of the commission to the earl of Bath in 1586,

> to save [reprieve] whom you shall think good to be saved, and to slay, destroy and put to execution of death such and as many [enemies and rebels] as you shall think meet by your good discretion to be put to death.

He possessed, therefore, truly vice-regal authority, the exercise of which depended, as always, on the particular personality of the holder of the office.[12] As far as I have been able to ascertain the Lord Lieutenant received neither fee nor even expenses, but in Tudor England all in positions of authority were expected to make whatever profit they could by virtue of their offices. All favours, especially the exercise of patronage in respect of appointments, required inducements in cash or in kind.

The cost of 'furnishing', that is equipping, the militia fell entirely on the counties. Every effort was made by the government to control the price of arms and armour, Bedford's instructions of 1570 quoting the following guidelines: a fully-furnished hackbut, 8s; a yew bow, 2s 8d, with arrows at 12d a sheaf; a pike 2s and a black bill 1s 4d. Armour cost rather more: that for a pikeman as much as 30s, more than a month's wages of a skilled workman. When away from home, either to attend a training camp or on active service, each man could claim 8d a day, even if it was a statutory holiday, for loss of earnings. To meet the bill the parishes were assessed by the JPs. In 1577 the Devon magistrates agreed that for the 'more contentation and ease of the people', that is the more well-to-do parishioners who would foot the bill, they would find it as far as possible out of parish 'stocks', that is funds intended primarily for the relief of the poor. In 1580 the Cornish JPs reported that twice-yearly training camps were costing individual parishes £4 to £5 a year, a sizeable burden. In 1587 at a time of dire emergency Sir Walter Raleigh suggested to Westminster that the Queen might find half the cost of the relatively small army of 2000 foot soldiers and 200 horse which he thought should be raised in Devon and Cornwall, but in the event the two counties received only the loan of some royal ordnance.[13]

An even greater difficulty, apparently, was the finding of what were called 'trainers'. In this case a chorus of moans from the counties had not gone unheard at Westminster and Bedford's instructions of 1570, after referring to a lack of 'sufficient gentlemen of knowledge', promised the appointment, in this case at the Queen's expense, of 'certain chosen honest captains', that is experienced professionals, to attend the camps and to give basic instruction. Some did materialise and a certain Captain Horde delighted the JPs of Cornwall by actually travelling around the county so that, not only could smaller camps be held and time and expense spared thereby, but the danger to public order of assembling large numbers of armed men could be avoided. In fact Captain Horde made himself so popular in Cornwall that in 1584 it was reported that some who hitherto were prepared to pay large sums to get exemption from service were actually keen to acquire more weapons at their own expense and to put in more practice.[14]

But such professionalism was rare, and for the most part it was a question of discovering volunteer gentlemen. In 1577 Cornwall appointed William Carnsew senior and three others, described as 'the best to be had', but by 1580 it was admitted that in spite of being assembled by such gentlemen twice a year,

> it is manifest that the soldiers profiteth little, as is seen in

> that they neither know readily how to put their match into their sock or take their mark to annoy the enemy, or [even] to stand soldierlike in their pieces.[15]

The inability, no doubt reinforced by disinclination, of the gentlemen is not entirely inexplicable. While all would have learned to ride a horse and use a sword as part of their education, few had any knowledge or experience of actual soldiering, especially the use of firearms. Unfortunately few probate inventories have survived for Devon but one of the earliest to include firearms is that of a rich widow of Uplyme, Alice Newall, who died in 1593 possessed of a musket, a caliver and other weapons. Altogether they were valued at only £4, but significantly her 20-ton bark was thought to be worth £60.[16] Sir Humphrey Gilbert's elaborate proposal in 1570 for the establishment of a military academy in London as a finishing school for young gentlemen, an alternative to that offered by the Inns of Court, fell on deaf ears. He himself, together with his younger half-brother, Walter Raleigh, was exceptional in his experience of military service, both on the continent and also in Ireland. However in a schedule of 'martial men' in a dozen English counties drawn up in June 1588 the county of Devon, with 16 names, actually heads the list, Cornwall being near the bottom with only four. The only Devon knight included was Sir Thomas Dennis who, with four others, was credited with service in Flanders. Four of those named had been to France and the greatest number, ten, had been blooded in Ireland.[17]

The limitations of the gentlemen as trainers of the militia extended, of course, to their role as officers. Quite apart from their lack of military experience, it is unlikely that any but those who were JPs, and then only a score or so of the really active magistrates, saw themselves as part of a county, let alone a regional, army. They felt happiest, as had many generations of their forbears, leading their immediate neighbours and their own tenants. However, in 1558 the earl of Bedford had begun to build on these existing territorial loyalties, endeavouring to extend them beyond the parish and even the hundred (the ancient grouping of parishes) to the 'division' of the county, borrowing the term used by the JPs in dividing their responsibilities. He prepared a schedule for Devon based on four divisions, one in the north of the county and three in the south, each not only containing a group of six to nine hundreds but each division made responsible for the defence of certain ports and potential landing places. Nothing so carefully devised is on record for Elizabeth's reign. To each division he appointed one of the leading knights, Sir John Chichester for the north, rather oddly Sir John St Leger for the east, Sir Thomas Dennis for the Exeter area and, as always, Sir Richard Edgcombe for the south west. It was an anticipation of his later deputies. Each divisional commander had a number of named 'Assistants' and also what he called 'Petty Captains', the difference being not one of military but of social status, all the 39 Assistants being knights and esquires and the 32 Petty Captains ordinary gentlemen. There were in all 10 knights in addition to the divisional commanders, but all but one were in the Exeter and east Devon divisions, near their places of residence. There were none in north Devon apart from Chichester. John Parker esquire was one of the Assistants. For the hinterland of Plymouth, besides Edgcombe there was only Sir Arthur Champernon over at Dartington. However, with no less than 75 militia officers, Devon was apparently well-provided with junior officers in the last year of Mary's reign.[18]

Thereafter, for over a decade, the records are silent, but some intelligence must have been relayed to Westminster for in 1562 the Privy Council wrote very sternly to the Devon JPs about one Simon Worth esquire. Seated at Worth in the parish of Washfield near Tiverton, he was one of Devon's many 'backwoods squires', as W. G. Hoskins calls them. News had reached their lordships that he had been appointed a militia leader although not only infirm but over sixty years of age.

> Their Lordships think it not meet that any man of those years or above the age of 45 or 50 at the most should be appointed to such a charge without some great and special consideration.

Fifty was, of course, a good age in Elizabethan England. Mr Worth died in 1563.[19]

By 1572 the Privy Council must have been demanding more information about the appointment of officers for a muster certificate for Devon returned in that year has appended to it a list of potential militia captains, 53 in all, 38 of them esquires and the rest plain gentlemen, but no knights. They included Bernard Drake of the east Devon family, John Amadas and William Hawkins of Plymouth, John Raleigh of East Budleigh and Adrian Gilbert and several other of the younger sons of local families who figure so largely in contemporary maritime affairs, including privateering. There was little militia service likely to be obtained from them. From Cornwall, however, that same year came a list of 22 of what were called 'Chieftains' and 58 'Captains', spread, no doubt according to residence, between the eight hundreds of the county (Stratton and Lesnewth being combined), a total of no less than eighty officers in all, considerably more in proportion to the able men, and indeed of the total population, than had been listed for Devon. The Cornish return is silent as to social status.[20]

Very soon after this the Devon JPs were endeavouring to persuade the Privy Council to agree, on the grounds of economy, to a trained county militia of only one thousand men, under five of what were now called 'chief captains', although they were also envisaging, if such could be found, the appointment of one captain for every hundred of their reserve force of ten thousand untrained men. (The overall totals fluctuate wildly from year to year.) By the end of the decade there begin to emerge the delineation, at least on paper, of what were called 'regiments', under the command of Italianate 'coronells'. In May 1584 each of the now five deputy lieutenants of Devon, all knights, was designated to command a regiment of 400 men, made up of roughly equal numbers of shot, bows and pikes. But to these were added four esquires, each with only 250 men. Clearly the status of its leaders dictated the size of each unit and there was no reference to any chain of command.[21] What, if any, had become the role of the squirearchy and gentry at large is veiled in mystery. It would seem that, perhaps due to the failing health of the earl of Bedford, his deputies were edging out all junior officers. Of the deputy-lieutenants Sir Arthur Bassett and Sir John Chichester conveniently lived in north Devon and Sir William Courtenay and Sir Robert Dennis near Exeter, but in the south with only Sir John Gilbert at Greenway there was no colonel 'seated' anywhere near Plymouth.

The death in June 1585 of the second earl of Bedford, although of late years a sick man and very rarely seen in the South West, created something of a power vacuum in the region, comparable with that of 1539. In addition to his Lord Lieutenancy he had also for some years been Lord Warden of the Stanneries and Chief Steward of the Duchy of Cornwall. With regard to his replacement, to some extent circumstances played into the Queen's hands in that the day before he died his eldest son was killed on the Scottish border and Bedford's heir was his 14-year old grandson. As was her wont the Queen slept on the problem and indeed only in February 1586 did she take the unusual step of confirming the appointment of the five deputies.[22] Less than a month later two of them, Bassett and Chichester, with a number of other leading gentlemen of Devon, died of gaol fever picked up at the Black Assizes at Exeter. Devon was now very bereft of military leadership.

But it cannot have been the resulting gap in north Devon which led the queen, later that year, to appoint as Lord Lieutenant of Devon William Bourchier, third earl of Bath, although it happens that his place of residence was Tawstock near Barnstaple. He was, in fact, the only adult nobleman with any roots in Devon, or indeed in Cornwall, but he was not put in charge of the latter county. Although not especially wealthy he and his widowed mother had recently rebuilt the family mansion, the splendid gatehouse of which still survives, bearing the date 1574. Aged only 29, he had no experience of county affairs and very little of soldiering except for a season in the Netherlands with the earl of Leicester. There was actually some family continuity not only in that his father had held the office of Lord Lieutenant for a short time early in Mary's reign, but also in that in 1583 in the church of St Mary Major in Exeter he had married Elizabeth, the earl Bedford's younger daughter. The city fathers had presented them with a basin and ewer of silver gilt and celebrated the occasion with a 'triumph' in Southernhay. Although they could hardly have anticipated events it was probably money well spent, ensuring that the city's claim, on account of its fairly new county borough status, to muster independently of the county would be respected. The Privy Council, very conscious of Bath's lack of experience, told him bluntly in October 1586 to be advised by his deputies, whose number was soon restored to five by the appointment of Hugh Fortescue esquire, of Filleigh, near South Molton, a descendant of the great Lancastrian lawyer, and Mr George Cary of Cockington near Torbay, the later Lord Deputy of Ireland.[23] Knights were few in Devon at this time, the newest, Francis Drake (later to be a deputy lieutenant), being fully occupied in the Queen's service at sea.

Cornwall's Lord Lieutenancy had already been bestowed on one of the few commoners to hold such an office in Elizabeth's reign, the recently-knighted Walter Raleigh. He had also succeeded Bedford as Lord Warden of the Stannaries and High Steward of the Duchy, both of which gave him some authority in Devon, of which he also became Vice-Admiral. At 31 he was slightly older than Bath but still unmarried. He had a respectable lineage, but so too had many others, including Sir Richard Grenville. His appointment was bound to be controversial and to one anonymous correspondent of Burghley's it was outrageous. After a swipe at those whose piratical activities gave Englishmen such a bad name abroad he proceeded to deplore in particular Raleigh's governorship of the tinners.

> Amongst so rough and mutinous a multitude, whose number we judge to be ten or twelve thousand, the most strong, able men of England, it were meet their governor were one of whom the most part did well account of, using some familiarity, and were abiding amongst them.[24]

The last point was very near the bone. Raleigh had no residence, or indeed any land to speak of, in the South West.[25] This, of course, the Queen herself could have remedied, as Henry VIII had done in the case of John Russell in 1539, just as she could have given him the Lord Lieutenancy of both counties, and even a peerage to add to his dignity. Perhaps she realised that he would never settle down so far from the Court and from the opportunities for the real power and influence which he coveted so much but never really attained. But even in his very infrequent visits he showed not only the capacity to devise a properly-structured militia combining the resources of both Devon and Cornwall but also an appreciation of the military topography of the peninsula unequalled by anyone actually living in the region. Plymouth, he argued, could more easily be relieved by the militia of Somerset than by that of Cornwall, the Tamar making the latter virtually an island.[26] Had the Spaniards landed in the South West in 1588 the failure to put Raleigh in complete charge of both counties might have been laid to Queen Elizabeth's charge as a ghastly mistake.

In the event Raleigh had few problems with the tinners, supporting them to the hilt in their claim to the right of mustering independently of the rest of the two counties. In March 1588 the earl of Bath complained that the deputy-Warden of the Stannaries for Devon, who happened to be Raleigh's elder brother, Carew, was failing to report on his quota of horsemen, but more revealing is Bath's further complaint that 'divers gentlemen of good accomplishment do now shroud themselves under the protection of the Stannary'. This they probably did with Raleigh's encouragement. It was the nearest approximation in Devon and Cornwall to a problem experienced in some other parts of the country, that of men claiming exemption from the militia on the grounds of their being retainers of great men.[27]

Bath and Raleigh had little time to settle into their new responsibilities. By early 1587, if not before, it was known that the day of reckoning was at hand. During the months preceding the actual arrival in the Channel of the Spanish Armada all Lords Lieutenants and their deputies, but especially those in the maritime counties, came under a barrage of inquiries and orders from Westminster. Had they seen to it that the militia captains had supplied 'dead and lame men's rooms', that is, filled up vacancies? Had they taken their captains to likely places of 'descent' by an enemy, 'to acquaint them with the ground'? Had they considered 'how to cover the soldiers from the enemy by the nature of the place [and had they provided] sconces, trenches or parapets'? To some of these the earl of Bath gave what were apparently satisfactory answers, though with regard to the making of sconces he replied that he lacked 'skilful persons' to instruct his pioneers and asked for assistance. Committees at Westminster, of which Sir Walter Raleigh was usually a prominent member, drew up their own schemes for the defence of the South West. One of their recommendations was that in the event of the militia having to withdraw in the face of the enemy the country should be 'driven', that is cleared of victuals. One is reminded of the 1940s, but one suggestion which I don't think even Winston Churchill ever contemplated was that the enemy should be 'kept waking with perpetual alarums'. On 31 March 1588 Bath, no doubt with some relish, replied to criticism that he appeared not to have provided 'carriages' for victuals, that Devon was quite 'unapt' for the use of carts or wains but that all parishes had been told to have nags ready for transporting the shot.[28]

As early as 7 December 1587, and just before Raleigh appeared at Exeter on a special mission to find out what was happening, Bath and his deputies reported that recent rumours of a Spanish fleet being at sea somewhere between England and Ireland had 'moved undoubted resolution in the minds of the people of this country, together with ourselves, to perform the parts of loyal subjects'. All, he added, was in a state of 'orderly readiness'. It is true that his army, at least on paper, had an orderly look about it. There were now three divisions, one in the north and two in the south. The north, Bath reported, had 1,217 trained men approximately equally divided between three of what were now simply called captains, Hugh Fortescue, one of the deputy-lieutenants, Hugh Pollard of King's Nympton and Anthony Monk of Potheridge, grandfather of the Civil War general. Each captain had a lieutenant, Fortescue's being William Stowford, one of those named in the list of 'martial men' as having seen service in Ireland. Each company had approximately the same number of shot, corslets (that is pikemen), bowmen, billmen and pioneers, and equal supplies of powder, match and bullets stored in Barnstaple, Torrington, Bideford and South Molton. In addition each captain also commanded nearly twice as many untrained men, also neatly categorised and described as 'furnished'. It is at this point that credulity becomes a little strained. The whole tableau was repeated for east Devon, although here there were only two captains, both of them Bath's deputies, and both knights, with correspondingly larger contingents. In south Devon, which stretched to the Cornish border and was therefore the only back-up for an as yet quite inadequately fortified and virtually ungarrisoned Plymouth, one large company was led by deputy lieutenant Sir John Gilbert and his brother Adrian and two smaller ones by the Champernon brothers, Richard and Arthur of the Modbury branch of the family, and by Thomas Fulford esquire of Dunsford. Mr George Cary of Cockington, although he too was a deputy lieutenant, seems not to have been entrusted with any command, except that he and Gilbert were jointly in charge of the Queen's ordnance which was stored at Ashburton. Why Mr Fulford was preferred to Mr Cary is not revealed though in view of Cary's subsequent quarrels, both with Fulford and with Gilbert, it is clear that there is a lot more behind Bath's dispositions than meets our eyes. Powder etc. for the east division was stored in Exeter, Tiverton and Cullompton and for the south in Totnes, Dartmouth, Plymouth, Tavistock and Plympton, corporate towns rather than gentlemen's residences being regarded as offering the best security. (But it was not all there. In 1589 the churchwardens of Chagford had in their custody 'four pounds of gunpowder in one treen [wooden] bottle [and] two pounds of match'.) In all three divisions men armed with the newer calivers had totally replaced the old hackbutters and there were now said to be 200 men in the county trained to use the even newer muskets. But even among the trained units bowmen were still nearly as numerous as pikemen though billmen were now in the minority, except in the untrained reserves. The splendid certificate sent from Devon early in 1588, beautifully engrossed on a large skin of parchment, may have reassured Westminster but we can only wonder what would have happened if, as we now know were not their orders, the Spaniards had attempted a landing in the South West. As late as 8 July, Sir Richard Grenville reported from Cornwall that eager as the gentlemen were to supply extra arms and armour there was none to be had for love nor money.[29]

The orders from Westminster to all the maritime counties were that all men should remain at home until ordered by some responsible person to assemble as instructed and march in an orderly manner to the coast. But who was to give the order? Raleigh wrote at some length to the Lords of the Council about the need to appoint a 'general of the Western Army' and although, following his visit to Exeter in December 1587 he wrote warmly to Lord Burghley about the loyalty of the earl of Bath, Sir Richard Grenville and Sir John Gilbert compared with that of others which he did not name, he was clearly worried that, in his own words,

> among many lieutenants [ten in all in the two counties] there be no straining of courtesy, lest by delay and confusion great inconvenience do grow to the country and advantage to the enemy.

In April 1588 Bath himself wrote from Tawstock explaining his difficulties in being so far from his deputies.[30] Exactly where he was during the last week of July is not on record. Let us not forget that he had had less than two years in which to learn his job. As yet he probably had a very hazy knowledge of the geography of the Channel coast. Did any units actually assemble and move to the coast? Sir John Gilbert later claimed, in the course of his quarrel with George Cary, that he had actually led a thousand men to the coast of Torbay, believing that, with its good 'road' for shipping, that was the likeliest place for a landing. Cary, he alleged, had skulked at home.[31] Otherwise nothing is on record in the state papers except in the flood of despatches from Howard and Drake at sea. Had a Spanish army landed, while it would certainly not have been welcome, I suspect that everyone would have stayed close enough at hand to protect his own house and family and waited for the earl of Leicester to come down with the Queen's army and do battle. It is surely only too likely that the large numbers of persons who, we are told, the Spaniards saw lining the clifftops of Cornwall and Devon, were there largely out of curiosity to watch the battle at sea and, of course, in the hope that a kindly Providence would send a crippled Spanish vessel ashore. For those in Torbay their vigilance was amply rewarded.[32]

Were the beacons actually fired? Was it, as Froude declared, by this means that England learned 'that the hour of its trial was come'? Repeated exhortations to the parishes, especially those on the coast, to keep their beacons watched by responsible men had always been accompanied by stern orders that none was to be fired except on the authority of the nearest JP, and they were few and far between in the coastal parishes, especially in the South Hams.[33] Moreover there is no evidence of any agreed code of signals in the South West such as was devised for Hampshire and the Isle of Wight. Finally it is clear from the state papers not only that on its initial sailing the Armada was first sighted by a fishing boat from Mousehole and on its second by Drake's ships sailing off the Scillies, but also that on both occasions the news was sent by messengers riding post horses. The post service seems to have been very efficient, though in 1596 the earl of Essex complained bitterly of the shortcomings in this respect of the city of Exeter.[34] Perhaps my questions can be resolved by those who are better acquainted with parish records.

The story of the Elizabethan militia did not end with the passing by of the Armada: it was just beginning. By the mid 1590s, with Spanish naval and military forces in Brittany the South West became even more vulnerable to invasion. In July 1595 a party of Spaniards actually landed in the far west of Cornwall and set fire to Mousehole and Penzance. When the frantic deputy-lieutenant, Sir Francis Godolphin, reached Penzance he found it being defended by 200 'naked', that is unarmed, men. Only the arrival of some of Drake's ships put the enemy to flight.[35] Sir Walter Raleigh, the county's titular commander-in-chief land forces, was in Guiana on the other side of the Atlantic, annoying the Spaniards in his own fashion. Since 1588 he had spent some time in Ireland and had also engaged in some very profitable privateering in home waters. He had also married and spent some months in the Tower for his indiscretion. When he was at home this was no further west than Sherborne.

In Devon the situation was somewhat different, if only in that the earl of Bath was usually at home at Tawstock and seems to have travelled quite frequently to Exeter to consult with his deputies. He found time in 1589, no doubt in the period of euphoria and relaxation following the disappearance of the Armada, to raise a delightful memorial to his mother, one much less extravagant than that erected by the altar in Tawstock church by his descendants to commemorate himself. He also found the time to take stock of his militia, in particular of his officers, all of them, as he explained to the Privy Council, appointed by his predecessor. He suggested a new arrangement whereby 'no captain have in his band above the number of 100 or 150 soldiers', a confirmation that the regiments of 1588 had lacked junior officers. The reply from on high was predictable:

> His Lordship should do well to appoint every knight that has charge to have the leading of 200, every esquire 150 and every gentleman 100.

The task of carrying out his plan involved him in a great deal of very difficult pacifying of those gentlemen of the old school such as Sir John Gilbert who feared the loss to their dignity of leading smaller 'private bands' as they were now called. Gilbert was doubly outraged as he happened to be away 'at the Baths' when the scheme was discussed by the deputy lieutenants.[36]

Our last glimpse of the earl of Bath must be on 25 November 1595. At Barnstaple Guildhall, as the Town Clerk records in his diary, accompanied by three north Devon JPs, Bath assembled all the parish constables of the north division,

> to give notice to those that were set to arms to be in readiness, and that the bills should be changed into pikes and the bows and arrows into muskets and calivers.[37]

Even among the 'North Devon savages', apparently, the hackbut had had its day.

It will not have been expected, from the title of my paper, that I would be arguing the importance of the militia as Elizabethan England's principal deterrent. It was her seapower, with some assistance from Providence in 1588, which saved the day. However some valuable lessons were slowly being learned, especially in the 1590s, both about the selection and training of officers and troops, and about the size and mobility of units. Above all the Lord Lieutenancy had become an established part of local administration. But the real interest of the quite considerable amount of documentation to which the Elizabethan militia gave rise lies in the light it throws on contemporary society.

NOTES

1. P[ublic] R[ecord] O[ffice], [State Papers Domestic Elizabeth] SP12, 131/21.
2. *The Devon Muster Roll for 1569*, ed. A.J. Howard and T.L. Stoate (1977); *Dorset Muster Rolls*, ed. T.L. Stoate (1978); *Certificate of Musters in the county of Somerset 1569*, ed. E. Green, Somerset Record Society, vol. 20 (1904), and *The Cornwall Muster Roll for 1569*, ed T.L. Stoate (1984).
3. Howard & Stoate, 38-9 and PRO, SP 12, 138/21.
4. PRO, SP 12, 134, fos 693, 697 and 97/1. The latter is undated and has been bound, in error, in the volume for 1574.
5. *Statutes of the Realm*, vols III, 832, and IV i, 317-8; and Howard & Stoate, 162.
6. PRO, SP 12, 215/16 and, for evidence of commutation, 188/36.
7. The standard work is Lindsay Boynton, *The Elizabethan Militia, 1558-1638* (1967). Also immensely useful is C.G. Cruikshank, *Elizabeth's Army* (2nd edition, 1966).
8. PRO, SP 12, 3/22; 2/31; 7/26-7; 6/61; 13/5; 13/18.
9. Howard & Stoate (for Devon) and Douch (for Cornwall), passim.
10. Devon Record Office, Quarter-Sessions Order Book 1592-1600, 104, 227.
11. G. Scott Thomson, *Lords Lieutenants in the Sixteenth Century* (1923), passim; PRO, SP 11, 12/53; SP 12, 18/36; 97/1 (misplaced); 74/34-36; 97/2.
12. Joyce A. Youings, 'The Council of the West', *T[ransactions of the] R[oyal] H[istorical] S[ociety]*, 5th series, x (1960); British Library, Harl. Mss, Cx, fo. 8.
13. PRO, SP 12, 114/1; 134, fo. 693; 140/22; 206/40, printed in E. Edwards, *Life of Sir Walter Raleigh* (1868), II, 36.
14. PRO, SP 12, 114/1; 169/5; 172/57.
15. PRO, SP 12, 112/22; 140/22.
16. M. Cash, *Devon Inventories of the Sixteenth and Seventeenth Centuries*, Devon and Cornwall Record Society, New Series, 11 (1966), 14. I owe this reference to Mr John Slate.
17. J. R. Hale, *Renaissance War Studies* (1983), 227-8; *H[istorical] M[anuscripts] C[ommission]*, 15th Report, Appendix v, 41.
18. PRO, SP 11, 12/67.

19. *Acts of the Privy Council*, ed. J. R. Dasent, vol. vii, 127-8 and J.L. Vivian, *Visitations of the county of Devon* (1895), 806.
20. PRO, SP 12, 89/26, 34, 41, 43-5.
21. PRO, SP 12, 91/26; 170/85.
22. PRO, SP 12, 176/50.
23. PRO, SP 12, 195/60-61; John Roberts, 'The Armada Lord Lieutenant: his family and career', *TDA*, 102 (1970), 79; SP 11, 9/15; Devon Record Office, St Mary Major parish register and S. Isacke, *Remarkable Antiquities of the City of Exeter* (1724), 137; *HMC. 15th Report*, Appendix v, 19.
24. PRO, SP 15, 29/126.
25. Joyce Youings, *Ralegh's Country: the South West of England in the reign of Queen Elizabeth I* (Raleigh, North Carolina, 1986), 1-7.
26. E. Edwards, op.cit., 112-17.
27. Youings, *Ralegh's Country*, 25-8; PRO, SP 12, 209/22; and Boynton, op.cit., 31-3, 98, 161-2, 167, 186-7.
28. PRO, SP 12, 204/58; 206/14; 209/42; 209/79.
29. PRO, SP 12, 206/14; 209/123; *The Churchwardens Accounts of Chagford 1480-1600*, ed. F.M. Osbourne (1979), 248; SP 12, 212/22,
30. PRO, SP 12, 206/40; 209/79 and 79i.
31. *HMC Salisbury Manuscripts*, vol. iv, 459.
32. PRO, SP 12, 217/10, 21, 22,
33. I owe this information to Mr Peter Cornford.
34. PRO, SP 12, 211, passim and 257/52.
35. A.L. Rowse, *Tudor Cornwall* (1941), 404.
36. PRO, SP 12, 230/107; *Acts of the Privy Council*, vol. xviii, 396; *HMC, Salisbury Mss*, vol. iv, 171. For these and other references to this 'little local difficulty', I am indebted to Mr John Slate.
37. J.R. Chanter, *Sketches of the Literary History of Barnstaple* (1866), 100.

The Defence of Cornwall in the Early Seventeenth Century

BY ANNE DUFFIN

This article will examine the question of defence as it related to Cornwall in the early seventeenth century, and particularly during the period from the beginning of the reign of King Charles I in 1625 until the outbreak of the English Civil War in 1642. During this period the security of the county weighed heavily on the minds of most Cornishmen. In 1625, war broke out between England and Spain, and by 1627 England was at war with France as well. France and Spain were the two major European powers and were almost always hostile to one another; to enter into war with both at once was extremely foolhardy, and made the threat of an invasion of England from the continent very real. One historian has remarked that the absurdity of the situation was as great as if Britain today found herself at war with Russia and the USA simultaneously![1] Cornish people (and indeed the inhabitants of all English maritime counties) were well aware of the invasion threat from both these powers, which made them all the more conscious of the inadequacies of their county's defences. Although both wars were over by the end of 1630, the threat of an invasion from France continued throughout the 1630s. However, during this decade the danger of a Spanish invasion was overshadowed by the threat posed by the Dutch, who by this time had both the means and the incentive to challenge English sovereignty in the Narrow Seas.[2]

The danger of an invasion from the continent was not the only threat to Cornish coastal security in the early seventeenth century. The inhabitants of the coastal towns were also faced with the constant threat from the Turkish pirates who were infesting the Cornish coast. (By 'Turks' were meant the pirates of many nationalities operating from the Barbary ports of Algiers, Tunis, Tripoli and Sallee.) Piracy presented an enormous problem to the English government: between 1616 and 1642 approximately 350-400 English ships and 6,500-7,000 English subjects were lost to the Barbary pirates. With few exceptions, almost all the ships which were lost were relatively small, lightly-manned merchant vessels from London or the west country ports.[3] Cornish people lived in constant fear of the Turks and it was not only those with maritime occupations who were at risk: the pirates often raided the coast as well. Besides creating an atmosphere of fear, Turkish piracy also had adverse economic consequences on Cornwall and the South West. In the first instance it caused economic decay by driving many merchants and fishermen to withdraw from their trade. For example in August 1625, the JPs of Cornwall told the Privy Council of the recent great losses of the inhabitants of East Looe, West Looe, Polperro and Fowey to the Turks. They had found 'the benefit of their fishing to be taken from them by which they are sustained', and feared even greater losses on the return of their shipping from Newfoundland.[4] Two days later, the Mayor and Brethren of Plymouth also mentioned the plight of the Cornish towns in their report to the Privy Council on the effects of piracy in Devon and Cornwall, and begged that

> some speedy course may be taken for the righting of these great grievances. Otherwise the merchant and mariner will be enforced to give over their adventures and trades which will tend to the great prejudice of His Majesty and the utter undoing of many in these western parts.[5]

Furthermore, the cost of redeeming the captives was high, and many towns were burdened with supporting the dependents of those taken captive by the Turks, and sometimes the victims themselves.[6] The Mayor of Liskeard included in his accounts for the year 1636 the following items:

> given to one Londoner that was spoiled by the Turks, 6d; given to a poor man that had his tongue cut out by the Turks, 12d; given to a poor man of Fowey that was spoiled by the Turks, 6d[7]

In July of the same year, the JPs of Cornwall petitioned the king to receive the advertisement of complaints recently received from the sea coast of Cornwall, and particularly from East and West Looe.

> About 2 months since, 3 barks of the said towns, on a fishing voyage upon the coasts, were taken by the Turks, and 27 persons carried away into miserable slavery, which loss falls

> the more heavily upon the said towns because of their former losses on two preceding years, wherein they lost 4 barks and 42 persons, whereby the said towns are not only impoverished, but by means of the wives and children of these poor captives (being more than 100 persons) are so surcharged, as they are likely to fall into great decay, and through the terror of that misery whereunto these persons are carried by these cruel infidels, the owners and seamen rather give over their trade than put their estates and persons into so great peril, there being now about 60 vessels and 200 seamen without employment. In other parts the Turks have taken other vessels, and chased others so that they have run on the rocks, choosing rather to lose their boats than their liberty. These Turks daily show themaselves at St Keverne, Mounts Bay and other places, that the poor fishermen are fearful, not only to go to the seas, but likewise lest these Turks should come on shore, and take them out of their houses.[8]

The high level of concern among Cornish people about the danger of an invasion from the continent, and the constant threat presented by the Turkish pirates, is reflected in the numerous petitions sent from Cornwall to the king and the Privy Council requesting naval protection for the county. Many petitions also pleaded for the erection of new coastal fortifications in certain exposed areas, and for the repair and improvement of existing ones. In 1625, the inhabitants of Fowey and Penzance petitioned the Deputy Lieutenants of Cornwall and the Council of War respectively, urgently requesting the fortification of their 'open and unfortified' towns.[9] The castles of Pendennis and St Mawes, which defended Falmouth harbour, and St Michael's Mount, were poorly supplied and in a dilapidated state. For example, in their correspondence with the government, the Governors and Lieutenant-Governors of Pendennis Castle expressed concern about the decay of the castle building itself, the shortage of ammunition and ordnance, arrears of pay, and the lack of essential food supplies, and gave as a particular reason for their concern the strategic importance of Falmouth harbour in the light of the threat posed by the pirates and the unsettled international situation. John Bonython, Lieutenant-Governor of Pendennis from 1614 to 1628, told Secretary Conway in November 1626 that

> the harbour I think is well known to be the best in the kingdom for a foreign enemy to attain unto; and for the safety of His Majesty's fleet outwards bound or for the expectance of an enemy to come for the Channel, they cannot lie better than at Falmouth; This fort, naturally so situated, that

> were it manned and victualled, is not to be taken. But being as it is, where we should have 40 pieces of ordnance we have none; no not one; nor almost any ammunition—our soldiers in great misery, having had no money this three years; the country disheartened in receiving of them, and seing the place so much neglected which is their strength the beat of Spaniards, Turks and Dunkirkers, all which I leave to your honours better consideration....[10]

The severe shortage of money and supplies at Pendennis in the 1620s made living conditions there appalling. For instance, in their petition of September 1625, the 'distressed soldiers of Pendennis Castle' informed the king that

> We can yet get no comfort, which other garrisons have received, whose wants cannot be greater, nor more miserable than ours, Your Majesty's poor petitioners, having been forced to pawn our bedding and other necessaries to buy bread to keep ourselves from starving[11]

However, the situation appears to have worsened, as in April 1626, Sir Robert Killigrew, the Governor of Pendennis Castle, petitioned the Council of War, asking

> that your Lordships would be pleased to take some course that the poor men there (being 50 in number) may have their pay, which they have wanted above 2 years, and are in that lamentable and miserable estate, that had they not lived on limpets (a poor kind of shellfish) without bread, or any other sustenance, with some small release from your petitioner, they had all been starved before this time.[12]

Still no money came to Pendennis, and in January 1627, Sir Robert Killingrew petitioned the Lords of the Privy Council that some of the soldiers had already died of malnutrition, despite the fact that he had wearied the Board with 69 petitions over the previous eleven years requesting the supply of the castle and pay for the men.[13] By the same token, from 1617 until the late 1620s, Captain Arthur Harris, Governor of St Michael's Mount, petitioned the Council for the repair of the castle, and supply for the Mount, and Francis Godolphin asked for money to improve the fort on St Mary's Island, Scilly, of which he was Governor. In February 1627, the king assigned £1,177 to Sir Robert Killigrew for repairs and improvements to Pendennis Castle, and £800 to Francis Godolphin for the same purpose on Scilly, and stipulated that the money was to be taken from the Loans of Cornwall.[14] However, the Petition of Cornwall to the House of Commons of April 1642 complained that this work was still uncompleted. It also said that St Mawes Castle was decayed, in that its platforms were rotten, its guns useless, its magazine empty and that the garrison at St Michael's Mount was in the same condition.[15]

There was also trouble afoot in the Cornish militia. From 1625, the Deputy Lieutenants in the South West, at the command of the Privy Council, had placed great emphasis upon the improvement and modernisation of the militia, to make it more effective. Measures included increasing the number of trained bands, mustering them, and training them more rigorously, whilst making all who had estates in Cornwall contribute arms. In August 1630, the Deputy Lieutenants of Cornwall told the Earl of Pembroke, their Lord Lieutenant, that

> we confidently assure your Lordship that the Militia or warlike provision of all sorts in this County is (as we conceive) in much readiness and well ordered to attend the worst adventure that may befall and more able to repel an enemy (with God's blessing) than in any time past.[16]

However this was not achieved without considerable opposition, even from amongst their own number. Bernard Grenvile of Stow remarked to Sir James Bagg in July 1630 that

> the Lieutenancy is grown into such contempt since the Parliament began as there be [those] that dare to countermand what we have on the lord commands willed to be done.[17] I have been a Deputy Lieutenant 2 or 3 and 30 years and never in all this time met so ill affections as now.[18]

Most gentry opposition appears to have taken the form of refusing to provide arms, or captains of regiments refusing to allow members of their regiment to attend musters. One incident cited by Grenvile occurred at his own muster at Bodmin, when Tristram Arscott presented him with a petition 'at so fit a time', according to Grenvile, 'to raise a mutiny if not a rebellion two regiments being in arms and in the midst of mustering'. Later that day, Arscott returned with a crowd of people and demanded an answer to his petition, which he claimed to have presented at the request of the local area.[19] The petition maintained that Cornwall was more charged with arms than any other county, and certainly more than Devon, Dorset, Somerset and Hampshire, and requested that the amount of arms demanded from Cornwall might be reduced to the level of 1586, or if not, that they might have contributions from other areas. It also suggested that the supply of arms should be in proportion to a man's estate, and that no man should be forced to muster at such a distance from his home that it involved him in great cost, or in travelling on the sabbath day.[20] Bernard Grenvile blamed the trouble on 'sundry ill dispositions poisoned by that Malevolent faction of Eliot'[21], and no doubt the factionalism which had pervaded Cornish politics in the 1620s had its part to play in the resistance to the modernisation of the militia in Cornwall. The 'Eliot' referred to by Grenvile was Sir John Eliot of Port Eliot in Cornwall, the well-known campaigner for parliamentary liberties, who played a large part in the passage of the Petition of Right in 1628, and in the Revolt of the House of Commons in 1629. However, this resistance must also be seen in the context of the impressment of Cornishmen for the expeditions to Cadiz and the Ile de Re, the billetting of riotous soldiers in Cornwall throughout much of the latter half of the 1620s and the resultant enforcement of martial law in the area. A complaint addressed to the Privy Council by the Deputy Lieutenants of Devon and Cornwall said that

> these said late employments have taken up so many as the very fishermen and sand bargemen have not escaped. The idlers which all countries willingly spare are either already pressed or have thrust themselves as voluntaries in the late expeditions. There wants nothing to consummate their undoing but the taking of their husbandmen.

As far as billetting was concerned, the main grievance among Cornish people appears to have been the government's apparent unwillingness to pay for the service. In the same letter, the Deputy Lieutenants complained that the existing supply of 2,000 soldiers had been billetted in Devon and Cornwall without a penny to maintain them, and that the king's honour was suffering by billetting soldiers without money.

> What, say the people, will His Majesty make war without provision of treasure, or must our country bear the charge for all England? Is it not enough that we undergo the trouble of the insolent soldiers in our houses, their robberies and other misdemeanours but that we must maintain them too at our own cost?[22]

These grievances were compounded in 1627 when the Deputy Lieutenants of Devon and Cornwall were issued with a commission to execute martial law upon the soldiers billetted in Plymouth and adjacent parts.[23] This commission was evidently enforced, as when some of the Deputy Lieutenants of Cornwall were called before the House of Commons in 1628 to answer questions about their behaviour during the elections of that year, they pleaded the need to delay, as they were about to execute some soldiers under martial law to make examples of them.[24] Martial law tended to evoke fears of an extension of central authority in the locality, especially as its use was not confined to soldiers, but was extended to civilians, to force them to take soldiers into their homes.[25] The imposition of martial law and the enforced billetting of troops, in addition to the burden of increased militia contributions from 1625, brought to a head in Cornwall and the South West opposition to the Deputy Lieutenants who were committted to enforcing the Council's policies for making the militia more effective.

Thus, during the late 1620s, Cornish people, like their contemporaries in the rest of the West Country, had to live with the threat of a possible invasion by either France or Spain, and with the presence and activities of the pirates who haunted their coastal waters. Despite the government's attempt to clear the coast of pirates, by sending out a naval squadron led by Sir Francis Stuart in 1625, the feeling remained in the West Country that they were receiving inadequate naval protection. Furthermore, the situation was not eased by the dilapidated state of the coastal fortifications in Cornwall. This too was reflected in other south-western counties. This feeling of insecurity had considerable repercussions on the national stage. One of those repercussions was that the competence of the Duke of Buckingham as Lord High Admiral of England was called into question, as it was his specific duty to see that the Narrow Seas were guarded. This point was raised in parliament, Sir John Eliot being one of its chief exponents, and was carried further in 1626 when one of the main charges brought against Buckingham in the Articles of Impeachment against him was his failure to guard the Narrow Seas.[26] However, although all parties at both local and national levels were equally keen to suppress the pirates and to defend the coast, their unity disappeared when asked to make a financial contribution to that end. There was great reluctance in the ports of Cornwall to pay ship money in 1626 to help furnish a fleet to fight the pirates. Many reasons were given for their inability to pay, including the demands by the king for the payment of a Forced Loan to help finance the wars, the heavy charge of billetting the soldiers, the decay of trade, plague, and the general poverty of the county.[27] However, there was also much resistance in Cornwall to the Forced Loan of 1626. For example, Sir John Eliot of Port Eliot and William Coryton of West Newton Ferrers were imprisoned for refusing to pay the Loan and for encouraging others to follow their example, whilst three other Cornish gentlemen were struck off the Commission of the Peace for the same offence.[28] Although to some extent this opposition was of a factional nature, much of the resistance to the Forced Loan in the county was evidently fuelled by the already heavy demands being made on Cornwall in terms of the additional militia contributions and the billetting of soldiers. In contrast, the collection of ship money in Cornwall in 1634 and 1639 was a resounding success for Charles I's government, Cornwall being one of the nine counties which paid the greatest proportion of their assessment.[29] Despite the poverty of the county, Cornwall apparently paid ship money willingly, at least until 1639. This reflects a completely different attitude to that adopted towards Charles I's fiscal expedients of the 1620s. Presumably, now that they were no longer burdened with impressment, billetting of soldiers, martial law and the Forced Loan which had accompanied the wars of the 1620s, Cornish people were prepared to give more credence to the king's statement of intent in the ship money writs that the money would be spent in taking speedy action against the pirates, and in safeguarding the country against the threat of invasion from the continent.[30] At last it seemed that the government was taking positive action to defend the Narrow Seas.

NOTES

1. C. Russell, 'Parliament and the king's finances', in C.Russell ed., *Origins of the English Civil War*, (1981), 104.
2. D.D.Hebb, 'The English Government and the problem of piracy, 1616-42', unpublished PhD thesis, London University, 1985, 288-310.
3. Hebb, 'Piracy', 192.
4. P[ublic] R[ecord] O[ffice], SP 16/5/32.
5. PRO, SP 16/5/36.
6. Hebb, 'Piracy', 192.
7. C[ornwall] R[ecord] O[ffice], Liskeard Mayors Accounts B/Lisk 288.
8. C[alendar of] S[tate] P[apers] D[omestic] 1636-7, 60-1.
9. Fowey: PRO, SP 16/11/78; Penzance: CSPD 1625-6, 207.
10. PRO, SP 16/40/29.
11. PRO, SP 16/6/137.
12 PRO, SP 16/25/105.
13. PRO, SP 16/49/6.
14. Pendennis: Acts of the Privy Council, Jan-Aug 1627, 64-5; Scilly: Acts of the Privy Council, Jan-Aug 1627, 64.
15. Mary Coate, *Cornwall in the Great Civil War and Interregnum, 1642-60* (Oxford, 1933), 22.
16. PRO, SP 16/33/111.

17. PRO, SP 16/147/46.
18. PRO, SP 16/147/14.
19. PRO, SP 16/147/46.
20. CSPD 1629-31, 27.
21. PRO, SP 16/147/14.
22. PRO, SP 16/88/46.
23. CSPD 1627-8, 440.
24. R.C. Johnson, M.F. Keeler, M.J. Cole, W.B. Bidwell (eds.) *Proceedings in Parliament, 1628*, 6 vols., New Haven, 1977–83, iii, 7.
25. C. Russell, *Parliaments and English politics, 1621-9* (Oxford, 1979), 337, 359.
26. Hebb, 'Piracy', 266-7.
27. PRO, SP 16/55/60.
28. These three gentlemen were Sir Richard Buller, Humphrey Nicoll and Nicholas Trefusis.
29. M.D. Gordon, 'The Collection of Ship Money in the reign of Charles I', *Transactions of the Royal Historical Society*, 3rd series, iv (1910), 142.
30. S.R. Gardiner (ed.), *Constitutional Documents of the Puritan Revolution, 1625-1660,* 3rd edn (Oxford, 1979), 105-6.

Defence and Security in the South West during the Commonwealth and Protectorate

BY IVAN ROOTS

Outbreak of civil war in 1642 directed the attention of both Crown and Parliament to the South West. Royalist sentiment was strong there but was by no means unalloyed. From the start there were conspicuous pockets of parliamentarianism, especially in urban centres like Plymouth and Exeter (in spite of, perhaps because of, the presence of the Cathedral) and even in the Duchy of Cornwall, Royalism prevailed initially over the region, as in Wales and the North, because from February 1642, when Charles I appears to have opted for a military solution to his politico-constitutional problems, he left the South East and the East of England *pro tem* to Parliament controlling London while he consolidated himself in more outlying areas. Distance from the capital was significant. Royalism might have been expressed openly and actively in Kent had not topography facilitated a Parliamentary ascendancy there, which was demonstrated by the quick suppression of a would-be cavalier rising in 1643. On the other hand, the limitations of, for instance, Cornish dedication to the king's cause can be discerned in the reluctance of ordinary Cornishmen to venture much beyond the Tamar. Provincialism was as powerful a pull during the war as it was in 1639 when Charles tried to rally national support against his rebellious Scottish subjects. A Dorset militia contingent had murdered their officer then rather than trudge northwards into alien parts. During the war itself the urge of troops to sit down in garrisons spoke of a home-loving inertia. There was neutrality, too, that could be jogged into the armed movement of the clubmen, particularly in the South West, as exasperation with the disruptions of daily life by the war grew. The disputable military and political control of the South West was important here.

In the second civil war the region, the mainland at least, was hardly directly involved. But it called for Parliament's wary eye then and throughout the Interregnum. Driven into exile, many royalists found their way direct to the Continent, but some took refuge in the Scillies and the Channel Islands. For a while they included the Prince of Wales himself and Edward Hyde, the future Chancellor and earl of Clarendon who started to plan on Jersey his great *History of the Rebellion.* The islands were seen not just as places of safety but as possible bases for a monarchical comeback. The threat this posed intensified following the execution of Charles I when the Rump felt itself insecure within the British Isles and for a while at least had cause to fear an onslaught by western European monarchies seeking to isolate and destroy the contagion of regicide.

The problems of defence and security, internal and external, facing the regimes of the 1650s were complex. On the surface they seemed those of patently illicit governments out to make themselves *de facto* against the *de jure* appeal of Stuart monarchy. Their difficulties could be considered peculiar to the 1650s. But there were also associated strains of a kind that harassed all legitimate seventeenth-century rulers lacking resources of money, information, adequate communications, and administrative expertise. The two sets of problems intertwined and could hardly be separated. Sheer military power and heavy policing might go some way towards resolving the one but could exacerbate the other, by underlining too thickly oppressiveness, pointing to a deficient confidence and thereby inhibiting a steady return to normality. Priorities in tackling them could never be worked up into one- two- or even five-year plans. Pragmaticism almost to the point of living from hand to mouth became *de rigueur.* The attention of historians has probably been more readily concentrated on the illicit side of things than was practicable for the governments themselves and one consequence may be that their actual achievements in 'healing and settling' by dealing with more routine demands were more considerable than might be supposed. But certainly for most of the period of the Commonwealth or (more pejoratively) the Rump (1649-53) establishing the reality of power was the prime pre-occupation.

The immediate foreign threat lessened as France and Spain, the most likely invaders, continued their part of the Thirty Years War, while facing internal revolts—the Portuguese, Catalans, the Frondes—and so far from contemplating intervention, joint or individual, looked for English neutrality or even hankered after securing the services of those New Model veterans, still in arms, whose reputation had spread throughout Europe. For their part the United Provinces would maintain their trading rivalry with

England but showed little disposition to interweave it in any positive way with military activity on behalf of the Stuarts, whose relationship to the House of Orange meant nothing to Dutch politicians unwilling to see an enhancement of the role of the Stadtholderate within the state. In 1649, at Burford, Cromwell cut the Levellers, who had taken no part in the king's trial and execution, to pieces and went on with great ruthlessness—not deplored by many Englishmen—to reduce Ireland to colonial obedience, and over the next couple of years to destroy the aspirations of a cynically covenanted King of Scots, the Prince of Wales. Victory at Worcester (3 September 1651) Oliver saw as 'a crowning mercy' and so, indeed, in some measure it was, but it would be a long time before either Scotland or Ireland could be considered utterly under control. Meantime other regions could not be ignored and so, like the Rump and the Protectorate themselves, we do well to glance towards the South West.

Royalists in the Scillies were not there for a quiet life. They meant to do as soon as possible what they could towards a comeback for their king and themselves into the mainland and in the interim to make things difficult for the usurpers and anyone actively or tacitly accepting their rule. In the identification of whom to assail they would not be over-scrupulous. So they preyed indiscriminately on shipping in the busy sea lanes around them, as privateers and even pirates—the line between the two was always a fine one. Miss Duffin has drawn attention *supra* to the perennial nuisance—it was often worse than that—of piracy along the south-western seaboards. Marauding provided a plausible excuse in the 1630s for the growing regularity of the issue of writs of ship money and their extension from obviously maritime counties to the whole country. Royalist privateers, claiming a legitimacy, plainly intensified the problem. Already committed to maintaining a land force in the South West under the experienced Major General John Desborough to check, repair and maintain coastal and inland fortifications and to co-ordinate garrisons and, as necessary, troop movements there, the Rump from 1649 onwards had to contemplate the far more difficult task (or so it seemed) of clearing out the islands, while still involved in the reduction of the larger units of the whole British Isles.

To clear the Scillies would call for combined amphibious operations at which hitherto the English appear not to have been very accomplished—one thinks of Buckingham's fiascos at Cadiz and the Isle of Rhé. Nevertheless, by 1651 disruption of shipping was becoming so acute, against the background of a possible royalist resurgence funded by booty, that by 1651 something had to be done. The Mayor of Exeter that year complained that 'Scilly rogues are so desperately bold that they stick not to steal vessels from under our very nose and almost within shot of our forts and castles'.

An unexpected intervention by the Dutch precipitated action. Tromp appeared with a powerful Fleet of St Mary's, ostensibly to protect their own traders coming up and down the Channel. This ploy could not be taken at face value. The Council of State were, it must be repeated, not unconcerned about the Stuart connexions of the House of Orange. Great trading rivals, the Dutch might also be thinking of occupying the islands for some as yet unglimpsed national purpose quite unconnected with dynastic altruism. The strategic nature of the location of the Scillies athwart the routes towards the Iberian peninsula, Western France, the Mediterranean and the burgeoning English colonies in North America was clear enough. The Dutch, whom the English had encountered traumatically at Amboyna in the East Indies in 1623, still remembered, still unavenged, were regarded as capable of anything. Fellow protestants they might be, some of them, anyway, but they could put profit before piety, Mammon before Jehovah. The Navigation Act of 1651 (and its successors extending well into the next century) together with the Dutch wars under the Commonwealth and Charles II reflect characteristic English attitudes. In what has been seen as an *odi-et-amo* relationship throughout the seventeenth century there was somewhat more hatred than affection.

So General-at-sea Robert Blake, lately victorious over Prince Rupert's fleet in the Mediterranean, was sent over to the Scillies to observe what was going on, prudently charged not to rush into a sea battle, but if it came to that, well, then it would have to come to it. As it happened Tromp was in no mood for a confrontation and after amicable discussions between the two commanders the Dutch fleet sailed away, leaving Blake in force on the spot. Here was an opportunity too good to miss for driving out the royalists and their hangers-on for good. There was reason to suppose that native Scillonians would be glad to see the interlopers go, leaving themselves to cultivate their own gardens and quietly fish their own waters. No resistance need be expected from them. Even so, to mount the operation would be difficult, especially in reducing the strong castle on St Mary's. But the seriousness and skill with which the Commonwealth forces tackled the invasion quickly dissipated the royalists' morale. After an initial show of readiness to stay put they sought honourable terms, to which Blake, a generous opponent, was responsive. They were allowed to sail away, many to slip quietly into obscurity at home, some to join up with royalists elsewhere in exile on the Continent or in the Channel Islands. Blake's victory would go a long way towards making the western waters safe for English (and incidentally other nations') shipping. There is irony in the fact that Blake's fleet owed much to improvements in the Royal Navy initiated by Charles I out of the proceeds of ship money. One may add that the way was left clear to extend to the Scillies the 'Domesday' Parliamentary survey of the lands of the Duchy of Cornwall.

With summer coming on the Channel Islands would no doubt have emerged as next on Blake's list of priorities. But as the situation in Scotland and the North of England deteriorated under the threat of neo-royalist armies, Desborough was sent up with a force to join Cromwell, Lambert and company. Blake, who like other admirals of the 16th and 17th centuries—post-Restoration Monck is an outstanding example—was capable of being both a general-at-sea and on land, was diverted to the mainland to look to the territorial integrity of the south-western region generally. Here his intimate local knowledge—he was born at Bridgwater and had been in action at Taunton, Lyme and Lostwithiel—was no doubt an advantage.

Commonwealth triumph at Worcester eased the crisis. Desborough came back and Blake was free to tackle the Channel Islands, a task of three-fold significance—to mop up royalists there, to free the sea-lanes of privateers and pirates, and to assert the English sovereignty which had been claimed and generally accepted for centuries. Here we can see clearly the inter-relations of long-term English national interests and the short-term concerns of an illegitimate régime struggling to survive by strengthening its control of the whole of the British Isles. The French who speak of 'les îles normandes'—they are after all nearer offshore to them than to England—had a perennial interest, though like the English they were ready for convenience to accept their military and commercial neutrality. It was not inconceivable that Charles Stuart might be prepared to barter them to Cardinal Mazarin in return for help towards his own restoration. But perhaps the stark fact of the islands as privateering and piratical bases weighed rather more. For that they were notorious even before the civil war. A Governor of Guernsey had reported that if (as was not unfeasible) the sea-rovers, who were capable from time to time of co-operation among themselves, made a determined move to take over the islands' fortifications—castles on Jersey and Guernsey—'it would be more nuisance than any Dunkirk in this kingdom'. (Dunkirk was still in the 1650s a busy privateering centre and it is not surprising that its capture should be a prime objective in Cromwell's later war with Spain, to give an English foothold on the continent and a thrust towards making safe the eastern English Channel and the North Sea.)

The pirates and privateers were a cosmopolitan lot—English, French, Dutch, Portuguese, even Moors who saw little point in trailing always back to the Barbary Coast after predatory actions. These men had no loyalties but to themselves, but they shared an immediate interest with royalists, who lavishly issued letters of marque, even ones with blanks to be filled in by recipients of whom no close enquiries were made. In general the native population were not directly involved in piratical or royalist activities. Their lives were led largely in agriculture, eked out with fishing and a little light industry. No doubt some economic advantage accrued from the interlopers' expenditure on provisions and services. But as in the Scillies Blake might well be confident that 'the people' would hardly stir to resist an invasion.

Before the 1640s the Royal Navy had visited occasionally to shew the flag and small garrisons were maintained in the forts built by the Tudors—Castles Cornet on Guernsey and Elizabeth on Jersey. In 1643 Parliament sent a naval squadron to assert its sovereignty and left a token force behind. But in the general pattern of the war the islands were inconspicuous and it was not difficult for the royalists swiftly to dislodge the Parliamentarians and to put their own increasing garrisons into strengthened fortifications, including a new one, appropriately called Charles. To drive them out would certainly be a tougher task for navy and army than it had been in the Scillies. Moreover, in October the weather could be expected to worsen, with sudden sharp autumnal storms. Landing soldiers and horses (sure to be seasick), guns and provisions would make demands that might not be met. To have to sail fruitlessly back to base at Weymouth would be an immense blow to Blake personally and to the Commonwealth's prestige. In the event commanders of both services, mariners and soldiers, responded effectively to every challenge, though it took some weeks to complete two operations, first on Guernsey, then on to Jersey.

The campaign is fairly well-documented, with at least a few detailed if partial accounts by participants. To clear Guernsey proved comparatively simple, but taking Jersey was more demanding, right from the tough landing at St Owen's Bay (23 October) with soldiers having to wade in waist-high from their open boats under fire from the shore. The smaller forts were soon eliminated, but Castle Elizabeth, as expected, put up a powerful resistance entailing a protracted siege. It was here that a large mortar, reputed to be a thirty-incher but in reality of about half that diameter, came into play. Its legendary size would probably have made it more dangerous to its operators than to the enemy, and though the real one, firing from one thousand yards under the supervision of Firemaster Thomas Wright, proved effective 'to the extreme terror and amazement' of the royalists, even that broke its mountings and twice suffered body fractures. It would have been unlikely to survive a third repair. That was not called for. But Castle Elizabeth did not capitulate until December. In victory Blake was as magnanimous as in the Scillies, allowing the Governor (Sir George Carteret, who had been there since 1643) and his troops and associates to leave. Carteret himself went to France.

A small military presence was left in the islands to mop up and to see that the islanders had 'protection to live peaceably at home and not [to] be troubled for anything done during the late war'. The royalists were permanently scotched, but sporadic piracy would remain an intractable problem for many more years to come. However, in comparative terms, it may be said, that between 1651 and 1659 the navy loyally kept the western seas open for the Commonwealth and the Protectorate. But as in 1659-60 the Good Old Cause began to collapse, commanders and seamen of the senior service, always more conservative than the army, turned first of all to demands for 'a free parliament' and then for government by king, lords and commons. Among the mariners who sailed under Admiral Edward Montagu, Pepys's patron, to bring Charles II home from Breda there must have been veterans of the Scillies and Channel Islands campaigns. (Blake had died in 1657).

After 1651 the main stress in security in the South West was on the mainland, though dangers from a long thinly-populated coastline and extensive sea lanes, intensified when war was declared on Spain in 1656, ensured that naval matters could not be forgotten. The Western Design in the Caribbean aroused echoes of the exploits of Drake and Hawkins nearly a century before. But the persistence of royalism at home, exposed to the urgings of activists, argued for the retention in centres like Exeter of 'regular' military forces which many in the government would have liked to get rid of—to save money, and to allow normality to sweep rather than merely creep back in. In defeat royalism was no more monolithic than in its civil-war heyday. Exiles who felt they had sacrificed all tended to be impatient with those at home who, like church-papists under Elizabeth I, wore their loyalty covertly to the point of a seeming total aversion to action. Some militants were prepared to co-operate with other dissident groups—presbyterians, levellers-turned-terrorists like Edward Sexby, author of the notorious tract *Killing No Murder* which encouraged assassination plots against Cromwell. (There were, in fact, several attempts, all inept, even ludicrously so). Other king's men, Simon-pure, held aloof. Organisations like the Sealed Knot tried to get co-ordination, with the desirability of joint action on a fully national scale in mind. But even the Knot was internally divided. Such factions can be traced right up to the exiled king's immediate entourage. His own mother was among those jockeying for advantage, offering diufferent strategies and tactics. Further complications came from infiltration by double agents, some ideologues, some hacks. Beyond that came the impact of government surveillance, particularly through the postal interception put on a systematic basis by Secretary of State John

Thurloe, 'father' of the G.P.O., whose 'state papers' published (most of them) in seven folio volumes in 1742, edited by Thomas Birch, are a prime source for many aspects of the Interregnum. Letters in code were mostly easily broken down and government always had a fair notion of what royalists hoped or even intended to do and were at any particular time doing. Even so, prudence dictated that 'the authorities' would always consider that something was going on that was leak-proof. That possibility and the political advantage that might accrue from crying up its dangers to those of the population, perhaps the bulk, who only wanted a quiet life, concentrated the mind of government. To exaggerate the threat was tempting, but it called for careful handling. Like the royal veto, such propaganda could be a wasting asset. In the long run rulers needed the appearance, in advance of the reality, of a growing acceptance by both the political nation and the population at large. Harping upon stubborn royalism pointed to an unstable base, deepening awareness of the abnormalities of the situation, lengthening the duration of 'the war period', stirring up old enmities, justifying the army's insistence on its own continuance as guarantor of the Good Old Cause—whatever that might have become—and bringing in its train political and social malaise, and financial strain, which would tauten when the Spanish war broke out.

It is against this sombre background that the republic's search for security and defence in the South West must be delineated. Distant from the seats of government, but strategically and economically vital, the region called for and got steady attention, indicated by the frequent presence there of John Desborough, member of the Rump's Council of State and later of the Protectorate's and of various commissions, including the Admiralty. His appointment became of major significance when overwhelming evidence was forthcoming late in 1654 and into 1655 of a royalist plot, concocted at home and abroad, for a concerted national uprising. The Lord Protector used the signs to sustain his onslaught on members of the addled first Proctectorate Parliament (September 1654—January 1654/5) for having remorselessly, though purged and warned, torn into the Instrument of Government, thereby giving shelter to 'weeds and briars' to grow in their shade. In sheer self-defence, therefore, the government argued, the parliament itself must be dismissed immediately upon its constitutional minimum of five (narrowly interpreted as lunar, i.e. 28-days) months.

The national rising did not come off. The agreed date (early March 1655) was anticipated country-wide by military alerts, searches, arrests and confiscations, forcing rapid changes of plans and rallying points, straining, indeed, breaking, communications between the leadership and the rank-and-file. In East Anglia there was not a flicker. In the North East hopes of the adhesion of Lord Fairfax, at once militarily experienced and locally politically influential, proved chimerical. A couple of hundred horsemen turned up at Marston Moor, but, finding no leadership, dispersed. Much the same happened right across the North. In the Midlands a football match and a race-meeting each remained just that. From Wales and along the line of the Marches, once the king's 'nursery of troops', silence. Only in the South West was there anything remotely dangerous and even that stirring rapidly took on a frenetic, desperate air.

The story has been told often. The leader was Mr John Penruddock, whence the traditional label Penniddock's Rising. He was a Compton Chamberlayne, Wiltshire, gentleman of moderate estate and (it seems) a hardly extreme political outlook, who organised hunting-parties and meetings over meals at the King's Arms, Salisbury, where local gentlemen, some of them former MPs and JPs, agreed to seize Winchester as their part of the general rebellion. The venue was changed at short notice to Salisbury where the assizes were in session—the plotters showed a strong aversion to 'loyalist' judges amenable to the Protectorate and to the sheriff, a notorious snapper-up of royalists' estates. There was also more than a hint of local political animosities as well as principles in all this. On the night of 11 March (three days tardy) 150 or so horsemen entered the city, pulled the judges and sheriff from their beds and taking them with them moved off west towards Bristol, which if it could be taken would be a jewel indeed in a crown for Charles II, whom they had proclaimed king. The force augmented itself to some 400 men but that was by no means as many as in optimistic forecasts. Even so, Desborough, then in London, was quickly sent down as 'Major General of (the six counties) of the West'. Before his arrival Penruddock's band had abandoned Bristol as an objective and turned off towards Cornwall, hoping for a more favourable reception there, or, if the worst came to the worst, for an escape route to the Continent. Numbers dropped away as they went *via* Cullompton and Tiverton and at South Molton, tired and depressed, they faced in a last stand, a contingent of regular troops marched up from headquarters at Exeter. At length they surrendered, according to Penruddock at his trial, on promise of generous and honourable terms. Swiftly repudiated, the claim was reiterated on the scaffold where Penruddock and a few others were exemplarily executed, to discourage others.

The attitude of the local populace excites curiosity. Clarendon (away from it all) would argue that if all the conspirators had been able to get their acts together, support would have been wide and rapidly expanding; *per contra*, Thurloe claimed that the more threatening the rising the more ordinary people would have turned out against it. But, then, each of these partisans would have said that, wouldn't they? In Thurloe's favour we hear of a volunteer force quickly raised in London in face of the national danger and of another deployed in Somerset to defend that county at least. That may have had something to do with Penruddock moving away from the road to Bristol. Analysis of prisoners taken at South Molton shows that they were chiefly gentlemen and 'meaner' men—tenants, domestic servants—dependent upon them. The unreadiness of the bulk of the people to stir encouraged the Council of State to continue to consider very seriously the re-establishment of local militia musters—a project which formed a part of the multi-motivated 'system' of the Major-Generals.

Penruddock's forlorn enterprise which did take off failed for much the same reasons as those which did not. Brave but incompetent, loyal to their king but shot through with jealousies, unable to overcome the inertia of all but a few (and among *them* fewer socially or politically top men, nationally or regionally), isolated from other disaffected groups, incapable of eliciting support from communities which had no inclination to have civil war renewed among them, they did not stand a chance. Government surveillance robbed them of the vital spark of surprise. Bad luck supervened. Even the weather—the South West then as now was often soggy—conspired against them, lacking as they did the royal sun of the presence of Charles to shine upon them. Yet one may doubt that a more concerted rebellion could ever have been a match for the veterans of the New Model Army. Penruddock's fate confirmed the view of some royalists that only a larger scale invasion, heavily armed, could be at all effective. For rather more, that would have been too high a price to pay, especially as they could identify no power altruistic enough to do the job and then quietly withdraw. Such men preferred to wait and see. In the upshot, they were right. Luck at last went their way in 1659-60—or was it Providence bringing about the disintegration of the Good Old Cause, now causes, some of them novel, and the startlingly decisive moves, unannounced in the stars, of that deep, dark Torrington man, George Monck?

That lay a long way off. The fiasco of the spring of 1655 was followed by the government's exploitation of the dangers, real or imaginary, hanging over state and society. The South West was significant here again. Based upon the appointment (already noticed) of Desborough as Major-General of the West, backed by local 'commissioners' named from the six counties, the so-called Cromwellian Major-Generals 'cantonised' England and Wales under detailed individual commissions and general instructions from the Council of State, which mingled indiscriminately the emergency requirements of a usurping régime and those standard to a *de jure* state with meagre resources. Control and suppression, though important, were only one objective. As indicated above the revival of trained bands was significant, too, making possible the reduction in numbers, politically and financially desirable, of regular troops in (theoretically) constant pay and, it was hoped, enlisting in a disciplined, overt way the co-operation of those natural rulers of the countryside—the gentry—as commissioners, equivalent to the former deputy lieutenants, and under the watchful eye of the Major-General as a sort of Lord-Lieutenant. Perhaps there was a naive optimism here, but the degree of positive support for or tacit acquiescence in the Protectorate has yet to be seriously assessed. The security of the regions—the South West perhaps particularly—against foreign invasion, renewal of civil war, rogues and vagabonds, Quakers, pirates and whomsoever, might be something for which people at all levels in the localities might be ready to summon a genuine loyalty. The implementation of a traditional social policy—the poor law, for instance—as a form of social control might well appeal to conservative gentlemen worried about the behaviour of the meaner sort of people, who from time to time, though never ungovernable and rarely enthusiastic in defiance of all authority, were not always as deferential to the élite as current social theory demanded.

The 'system' of the Major-Generals was too abruptly introduced and as incontinently abandoned by Cromwell, never enthusiastic for it, responding to parliamentary criticism, for us or for contemporaries to estimate what its long-term consequences might have been. It was, also, too ambivalent—at once centralising and localistic. That contemporary reference to 'cantonising' is not without significance. Moreover, the more oppressive elements, including the decimation tax upon royalists and restrictions on their freedom of movement, were 'necessarily' (in the government's judgment) uppermost to start with. Again, individual Major-Generals, a few of them unpleasant bigots, though no doubt sincere, gave an unwelcome priority to some of the least appealing of their instructions, of which the most egregious was 'moral reformation'.

So early in 1657 the major-generals were bundled quickly away. What happened afterwards in the South West goes beyond the remit of this article. But that there was continuity mingled with change in the problems and possibilities of security is plain. The war with Spain may have been mainly fought at the eastern end of the Channel and on land, but it was not without impact on the South West. Exeter, for instance, traded heavily with the Low Countries and so was not indifferent to the fate of Dunkirk privateers nor to the war's impact upon Spanish sea-power, which has surely been underrated by historians bemused by the questionable onset of the decline of Spain. The death of Cromwell (3 September 1658) long expected to be the signal for the collapse of the Protectorate, found royalists in the South

West as unready to rise again as they were elsewhere. Richard Cromwell succeeded without opposition, indeed, with fulsome acclamations, some no doubt feigned, others quite possibly genuine in their sentiments, from this region and others. At Richard's resignation (April 1659) again silence. Booth's rising in Cheshire (later in the year), which significantly attracted little local backing or reinforcement, was not emulated in the South West. But certainly before Monck revealed his intention to bring back Charles II, there were signs that communities down here were ready to obey 'a free parliament', in whatever decisions it might come to, in order to stem a drift into what looked like anarchy.

After the Restoration security remained a problem, now for a *de jure* government. Even before the euphoria of the early 1660s wore off there were circumstantial rumours of plots by protestant sectaries in centres like Tiverton, Axminster, Dartmouth and Exeter. Fears of a Baptist rising at Plymouth in the autumn of 1662 saw the county militia called out and a force of 'redcoats' raised in Exeter. The presence of prominent Interregnum figures imprisoned in the South West (Lambert in Guernsey, Henry Vane—until his trial and execution—Wildman in the Scillies, James Harrington on St Nicholas Island, off Plymouth) prompted local authorities to be wary of disbanded New Model men in their midst, though most of them seemed happy enough to fade away rather than to die for any cause, good and old, or new and better. But dissent, splattered with revolutionary mud drying out into dust, survived in the South West and informers made a good thing out of its infringement of the Clarendon Code.

Finally, there would be two risings, each stemming from an invasion, down in the South West in the closing decades of the century. In the first, Monmouth's (1685), it was reported that the sea-green colours of the Levellers were worn here and there again. It attracted little or no local gentry support and failed signally, though not before it had taught James II not to rely upon the militia. 'No standing armies', though, continued to be a powerful slogan to men whose memories went back to '41 or '49 or '59. The second, William of Orange's landing at Torbay with a considerable Dutch force having evaded the Royal Navy, was a better-calculated risk, less romantic and, of course, successful in leading to that not unmixed blessing, the Glorious Revolution. The reasons for its victory are complex and would take us way beyond the region, but the difficulties of ensuring from London a total control in town and country of the South West are certainly among them.

FURTHER READING

The fullest general narrative of these years is still S.R.Gardiner's *History of the Commonwealth and Protectorate*, 4 vols (1903, new Edition). The context is partially provided in more recent works: B. Worden, *The Rump Parliament* (Cambridge 1974) and A.H. Woolrych, *Commonwealth to Protectorate* (Oxford, 1981) and, more broadly, I. Roots, *The Great Rebellion* (1966 and later editions).

For the maritime aspects see J.R. Powell, *Robert Blake, General-at-Sea* (1977); A.G. Jamieson, *A People of the Sea: The Maritime History of the Channel Islands* (1986); G.R. Balleine, *All for the King: the Life of Sir George Carteret* (Jersey, 1976); M.F.H. Ellis, 'The Channel Islands and the Great Rebellion', *Bulletin of La Société Jersiaise*, xiii (1937); and S. Bull, 'Mortars at Elizabeth Castle Jersey, in 1651'; *Fort* (*The International Journal of Fortification and Military Architecture*), xiii (1985).

On the royalists generally see D. Underdown, *Royalist Conspiracy in England* (New Haven, 1960) and P.H. Hardacre, *The Royalists during the Puritan Revolution* (The Hague, 1956). On Penruddock see W.N. Ravenhill's articles in *The Wilts. Archaeological and Natural History Magazine*, xiii-xv, (1872-5) and A.H. Woolrych, *Penruddock's Rising 1655*, a Historical Association pamphlet (1955 and later editions). For Desborough see articles in *The Dictionary of National Biography* and R.L. Greaves and R. Zaller (eds.), *Biographical Dictionary of British Radicals in the 17th Century*, Vol. I (Brighton, 1982), which also contains a sketch of Robert Blake. See also D. Underdown, *Somerset in the Civil War and Interregnum* (Newton Abbot, 1973) and M.P. Ashley, *Cromwell's Generals* (1954).

The Major-Generals are surveyed in D.W. Rannie, 'Cromwell's Major-Generals', *English Historical Review*, x (1898) and I. Roots, 'Swordsmen and Decimators' in R.H. Parry (ed.) *The English Civil War and After* (1972). The instructions to the Major Generals are to be found in context in W.C. Abbott (ed.) *The Writings and Speeches of Oliver Cromwell*, Vol. III, (Cambridge, Mass. 1945), which also includes a specimen of a commission for a Major-General, William Boteler). Cromwell's declaration explaining the introduction of the Major Generals is printed in *The (Old) Parliamentary History*, xx (1765). Pertinent, too, are Oliver's Speeches of 22 January 1654/5 and 17 September 1656, to be found in Abbott; in any edition of T. Carlyle's *Letters and Speeches of Oliver Cromwell*, (1846 etc.); and in I. Roots (ed), *Speeches of Oliver Cromwell* (forthcoming). D. Underdown, 'Settlement in the Counties 1653- 1660' in G.E. Aylmer (ed.) *The Interregnum: The Quest for Settlement* (1972), A. Fletcher, *Reform in the Provinces* (New Haven, 1986) and his 'Oliver Cromwell and the Localities: The Problem of Consent' in *Politics and People in Revolutionary England: Essays in Honour of Ivan Roots* eds. C. Jones, M. Newitt and S. Roberts (Oxford, 1986) are also useful.

The post-Restoration conspiracies are explored in R. L. Greaves, *Deliver us from Evil: The Radical Underground in Britain 1660-1663* (Oxford, 1986). Two relevant unpublished dissertations are P. Jackson 'Nonconformity, Devon, 1649-1689' (Exeter 1987) and N.R.R. Fisher, 'The Deputy Lieutenants in Somerset 1660-1667' (Graduate Diploma in Humanities, Middlesex Polytechnic, 1987).

The West Country-Newfoundland Fishery and the Manning of the Royal Navy

BY DAVID J. STARKEY

The defence and security of sea-borne commerce was of vital concern to merchants and strategists intent on promoting overseas trade and colonisation in the seventeenth and eighteenth centuries. The prevailing mercantilist orthodoxy had it that commercial growth and political power were inextricably and causally linked. While an expansion in trade depended upon the exercise of political and military power to establish and maintain advantages over rivals, commercial prowess, in its turn, was at the base of political might, for it created the wealth necessary to sustain expansionist policies. Moreover a flourishing overseas trade, it was argued, had a further utility in that it served as a vital training ground for mariners who could be recruited into the Royal Navy when the security of the empire was threatened. The belief that the mercantile marine acted as a crucial 'nursery of seamen' was widely held by contemporary merchants and governments alike and was an important and persistent factor in the long-running debate on the role of trade in international affairs. It was an argument propounded by advocates of most sea-borne trades of the period—from the West India trade to privateering to the coastal coal trade—but it was particularly pervasive in relation to the oldest of Britain's Atlantic staple trades, the fishery conducted in the waters off the island of Newfoundland. By examining the contention that the Newfoundland fishery was 'one great nursery of seamen and a principal basis of the maritime power of England' (William Pitt, 1760), it is hoped that some light will be shed on the complex relationship that commercial interest groups shared with the state and its naval forces.

I

The strength of the 'nursery of seamen' thesis in contemporary discussions and legislation regarding the Newfoundland fishing industry lay in the nature and operation of the trade. The commodity at the root of this business is properly known as *Gadus callarias Linnaeus*, or, as it is more commonly known, the cod. This is a high protein fish, the 'beef of the sea', its flesh being a valuable source of food, and its liver a source of 'train' or 'cod liver oil'. For a variety of reasons, cod flourish in the waters covering the submerged continental shelf off the seaboard of north-eastern America. Europeans were drawn to this plentiful supply of food from the late fifteenth century onwards, the fisheries of Spain and Portugal dominating the trade in its early stages. However, towards the end of the sixteenth century, the Iberian fishery succumbed to the attacks and encroachments of English and French fishing interests, and there developed a prolonged, often violent, Anglo-French struggle for control of the fishery and settlement of the island.[1]

From its beginning, the English cod-fishing trade was based in the West Country, initially due to the favourable geographical position of the Western counties, and subsequently due to the political tenacity and commercial expertise of the interest groups which developed in ports such as Poole, Dartmouth, Exeter and Plymouth.[2] It was essentially a seasonal trade, involving the annual migration of large numbers, often many thousands, of fishermen from the South West to the island of Newfoundland for the fishing season. Typically, the fishing fleets departed the West Country in the spring, crossed the Atlantic and established bases, or fishing 'rooms', on the Newfoundland shoreline where the cod were dried prior to their delivery in the autumn to the markets of Catholic Europe, principally Spain, Portugal and Italy. Though the English attempted to plant settlements on the island from the early seventeenth century[3] and though some fishermen—usually the desperate or the unfortunate—chose to winter on Newfoundland, the fishery remained principally a migratory business. Quite simply, until improving political and economic conditions made settlement a more attractive proposition from the 1750s onwards, life on Newfoundland was so bleak, hard and uncertain that most fishermen considered the long and hazardous annual return voyage preferable to settling there. Thus, in contrast to the experience of New England where settlement, and a sedentary fishery, developed relatively swiftly, the Newfoundland trade was conducted by merchants and fishermen resident in England, generally in the West Country, for much of the year.

In the context of the fiercely competitive commercial and political world of the seventeenth and eighteenth centuries, the international extent and migratory pattern of the English cod-fishing industry made it the object of much mercantile and governmental concern. In general, the character of the trade appealed to most commentators, for it fitted perfectly into the mercantilist conception of commerce and enterprise, having three great attributes according to contemporary strategists. In the first place, it was viewed as an 'export-orientated' trade. As very little cod was consumed at home, the main markets being in southern Europe, it was deemed to be an important earner of bullion and produce from potential competitors. Secondly, it provided employment for thousands of English people, for in addition to those who fished the Newfoundland waters and those who navigated the fishing vesels, there were the countless others engaged in the handicraft industries of the South West which supplied the equipment, the victuals and the shipping for the trade. It also generated profits for the 'Western Adventurers', the merchants in control of the business, profits which permitted further investment and employment in the trade. Thus, the cod trade created income and employment which sustained many ports and coastal villages of Devon and Dorset.

The third great attribute of the cod-fishing industry—the one which outweighed all others in the minds of politicians—was the belief in the fishery's value as a training ground for mariners. This was a powerful argument, for the structure of the Newfoundland trade meant that, potentially, it could help supply the Navy's seemingly insatiable demand for manpower in times of emergency. The fishery employed large numbers of men; in the seventeenth century a good season might see upwards of 10,000 men leave the West Country for the fishing grounds, while in the 1770s over 20,000 undertook this annual pilgrimage. The migratory basis of the trade meant that nearly all the fishermen and mariners were available for much of the year. It was known as a clean trade, in that it was free from tropical diseases and other complaints which traditionally afflicted the sailor. Moreover, and most importantly, the cod-fishing industry employed and relied upon skilled and experienced seamen and fishermen, for high standards of seamanship were required and learned in this dangerous long-distance trade.

Thus, ran the argument, trade and defence were intimately linked: the West Country-Newfoundland fishery was an ideal 'nursery of seamen', providing the Admiralty with a plentiful, an available, a healthy and a skilled reservoir of potential naval recruits. While its appeal was widespread and its logic rarely challenged, a closer examination of this thesis reveals the self-interest and naivety of its main proponents, and the distorting effect it had in policy towards the incipient colony of Newfoundland and towards the fishery itself.

II

The 'Western Adventurers' were a pessimistic body of men. Engaged in a vulnerable and volatile industry they were quick to complain to the Board of Trade when their interests were threatened, whether by enemy or piratical action, by the incursions of foreign fishermen, or by settlement in Newfoundland which might undermine the migratory basis of the trade. A regular feature of the numerous petitions emanating from the cod-fishing ports of the South West was a reminder of the wider benefits of the activity. 'The fishery', it was stated, '...is a true prop of the trade and navigation of this island'. Moreover, 'a fishing trade is one great and certain nursery of seamen.'[4]

Such noble statements stress the national service that the fishing merchants insisted their enterprise performed. The onset of war, however, exposed the hollowness and propagandist nature of these assertions. On such occasions—and they were frequent in the seventeenth and eighteenth centuries—the Adventurers were placed in something of a dilemma. For, having assiduously cultivated the idea of Newfoundland as a 'nursery of seamen' whenever they had occasion to raise complaints about the trade, they could hardly object openly if, in wartime, the Navy claimed large numbers of men in the coastal towns of Devon and Dorset, thereby crippling the fishery. Instead, they either accepted the situation or else attempted to use their influence, nationally and locally, to avoid or resist the Navy's customary form of wartime recruitment, the press gang. Many chose the latter course of action, using whatever tactics, legal or otherwise, at their disposal. Immediately they heard of an impending press, the merchants speeded up their departures, hoping to get their vessels away before the Navy arrived. Protections from the press were frequently sought in the south-western ports, often by local MP's eager to please their constituents and patrons. Thus, in 1667 Sir John Colleton boosted his election chances at Dartmouth by declaring that he had obtained more protections from the Duke of York than any man alive. Bribery—or the payment of 'charges and gratulation'—was a common resort of the fishing merchants, and it seems that more than a few Admiralty officials were open to this type of persuasion.

The more direct methods of avoiding the press employed by the seamen and fishermen themselves were often openly or tacitly supported by the fishing merchants. At Bideford in 1653, for instance, the men banded together, armed themselves and warned the press master to take them at his peril. The local magistrates—all in the Newfoundland trade—connived openly at this and did all that they could to obstruct the gangs. The mariners of Poole gained a reputation for being particularly resistant to the press; armed with guns, they hid in the quarries of Purbeck or on board sloops in outlying caves until the press gang departed. If they fought a naval cutter, the support of local juries was assured and a naval captain who killed a Poole fisherman during a scuffle was found guilty of murder.[5]

How effective these measures proved in frustrating the press is difficult to assess owing to the paucity of accurate data on the numbers recruited into the Navy during the period. It would seem likely, however, that these obstructive actions were token gestures, born out of the frustration felt by the fishing merchants in wartime. Whether or not the Newfoundland fishery actually provided a large number of trained seamen for the Navy, the idea that it *should* do so, as enunciated by the Adventurers themselves when it suited their needs, attracted the attentions of the press and created frequent and chronic labour shortages. Those men who were not themselves pressed away, promptly fled inland to hide, or stayed in Newfoundland at the end of the fishing season. Their destination made little difference to the fishing merchants whose business was heavily labour-intensive and any manpower shortages, particularly of skilled men, inevitably disrupted the activity. Thus, the trade was regularly and severely depressed by the onset of war—from 1625 through to 1793—as the naval press scattered, by recruitment or fear, the essential labour supply.

III

The disasters inflicted upon the Newfoundland fishery in wartime indicate the insincerity and flawed logic of the Western Adventurers' adherence to the 'nursery of seamen' thesis. This being so, why did this rather superficial dictum form such a prominent weapon in the fishing merchants' propaganda arsenal and why did successive governments find it such an appealing argument?

Clearly, in their propagation of the idea that their trade was a major source of the nation's maritime strength, the fishing merchants sought preferential treatment for their industry and advantages for themselves within the trade. This was particularly true in the early seventeenth century when the Adventurers perceived a threat to their interests from the early settlements established on the island of Newfoundland. Though these plantations were short-lived—largely owing to the inhospitable climatic conditions on the island—they represented a potential source of competition to the West Country's migratory trade. This rivalry was not confined to the catching and selling of fish. It extended to the shorelines on which the fish was dried, and in this respect the settlers held a clear advantage over the migrants.

The development of a sedentary fishery, therefore, threatened to encroach upon the 'free trade' practised by the fishing merchants resident in the mother country, as had been the case in New England.

In this context the Western Adventurers demanded that their rights to continue a 'free' migratory fishery—which would guarantee a supply of trained seamen—should be confirmed. The 'Western Charter' of 1634, re-enacted in 1670 and 1675, recognised these rights, enshrining the basic elements of Western control of the fishery. The charter stated that,

> no planter shall cut down any wood or plant within six miles of sea shore .... no inhabitant or planter shall take up best stages before arrival of fishermen .... no fishermen or seamen shall remain behind after fishing is ended

and enjoined,

> that Admiral, Vice-Admiral and Rear Admiral do put these orders in execution and preserve the peace .... and bring offenders for any crime to England.[6]

These regulations consolidated the hold that the Western Adventurers had established over the trade and placed settlers on Newfoundland in a subservient position, both in the fishery, and in the control of affairs on the island itself. Planters were not allowed to settle within six miles of the coast, thereby reserving the shoreline for the drying of fish, nor were they allowed to claim the best fishing rooms before the arrival of the transient fishermen. Power on the island was vested in the Admirals, whose sole qualification was the fact that they arrived first each year in a particular bay or harbour, and whose jurisdiction covered the 'preserving of the peace' in that locality. They had the right to dispense justice as they thought fit in most matters, though serious crimes were to be judged in the ports of Western England. In effect, the government, convinced of the utility of the migratory fishery as a training ground for mariners, gave the Western fishing merchants exactly what they wanted—direct control of the fishery and a dominant influence on the growth of settlement, and potential competition, in Newfoundland.

The granting of such draconian powers to a vested interest group inevitably had consequences for the development of the island, and the Western Adventurers have traditionally been viewed as the 'villains' of Newfoundland's early history, being largely responsible for its prolonged failure to grow in the form of the mainland American colonies.[7] This is only partly true, however, for the Charter of 1634 reflected the concerns of the fishing merchants in the 1630s, a time when Sir David Kirke was leading an aggressive attempt to establish and govern a plantation on the island. The failure of this threat was partially due to the overt hostility of the migrant fishermen but thereafter, there is little evidence to suggest that the Western Adventurers deliberately obstructed settlement. The conditions of the 1634 Charter were very difficult to impose. Winters on Newfoundland were hard enough to deter settlers, and those small settlements which existed had little in the way of secondary employments to sustain them outside the fishing season. The French remained a serious threat to the island's planters until 1763. Moreover, the fishing merchants themselves had reason to encourage settlement for they controlled the shipping involved in this trans-Atlantic trade, and were undoubtedly aware of the possibilities of diversifying their interests into supplying this isolated colony with much needed commodities.

In the event it was not until the mid-eighteenth century that improved economic and political conditions permitted the expansion of settlement on Newfoundland. A series of good fishing seasons, the development of seal hunting and shipbuilding as winter employments, and above all, the defeat of the French in 1763, lifted prospects for settlers and encouraged more to reside permanently on the island. These developments signalled the end of the migratory fishery, though, ironically, it flourished in its dying years as the market conditions of the 1770s and 1780s proved especially advantageous to cod producers. The fishing merchants anticipated the demise of their migratory trade and found it convenient and profitable to adapt to changing conditions by transporting and supplying the colonists, buying their fish and carrying it to market. Indeed, many merchants and fishermen transferred themselves and their businesses to the island. During this long process of development and re-adjustment, the Western Adventurers continued to plead their case to the state when the need arose. The response of successive governments was remarkably consistent from the 1630s through to the 1770s. Despite changing modes of operation within the fishery which led to the employment of fewer, though more productive, men, and despite the evident growth of population on Newfoundland from the 1740s, government policy remained based upon the seventeenth-century belief that the migratory fishery should be encouraged to foster the nation's naval manpower. In 1699, the Newfoundland Act reiterated the terms and conditions of the 1634 Charter, though naval officers were given some authority in the administration of the island's affairs. Successive Board of Trade reports and proposals in the eighteenth century confirmed this increasingly inappropriate policy, generally in the face of opposition from the Western Adventurers. In 1775, the government's ignorance of changing conditions within the fishery reached its height with the passing of Palliser's Act, so called after the naval officer who framed the bill. This anachronistic and unrealistic piece of legislation attempted to turn the clock back over a hundred years and proved to be totally inappropriate to the needs of the island and the fishing industry, and therefore wholly ineffective. The preamble to the Act summarised the state's view of the Newfoundland trade as follows:

> Whereas the fisheries ... have been found to be the best nurseries for able and experienced seamen, always ready to man the Royal Navy when occasions require ... it is therefore of the highest national importance to give all due encouragement to the said fisheries, and to endeavour to secure the *annual return* of the fishermen, sailors and others employed therein to the ports of Great Britain at the end of any fishing season.[8]

Thus the 'nursery of seamen' argument remained the bedrock of the government's Newfoundland policy until the closing stages of the migratory industry, despite its irrelevance to conditions in the fishery and the open opposition of the Western Adventurers. When the trade finally collapsed in the early 1790s, one of the West Country's most important maritime activities ceased. Yet there were relatively few complaints from those engaged in the trade for the majority had responded to changing conditions in the colony and adapted their interests accordingly. Politicians, on the other hand, failed to appreciate the implications of the island's development from the mid-eighteenth century, a fact reflected in their reactionary and unrealistic policies.

IV

It is impossible to test directly the validity of the 'nursery of seamen' argument for there is no information available on the number of naval recruits who learned their seafaring skills in the Newfoundland trade. However, it is highly unlikely that the fishery formed the mainstay of the nation's maritime power, as the fishing merchants repeatedly asserted. They used this argument as a fairly shallow piece of propaganda to win for themselves significant advantages in the control and organisation of the trade. Governments, ignorant of the structure of the trade and ever obsessed with the problem of manning the Navy in times of emergency, were very susceptible to this argument—in the end, to an embarrassing degree. In reality, though the Newfoundland trade was frequently cited as an essential prop of Britain's maritime strength, it would seem that the opposite was true. The Newfoundland fishery, and the island itself, was very much dependent upon the power of the Navy. As with the other colonial staple trades, the cod fishery was a long-distance trade, vulnerable to attack—both physically and commercially—in a competitive and violent age. It was therefore only secure, and only able to flourish, when the state and its Navy could establish and maintain a favourable imperial framework.

NOTES

1. See Harold A. Innis, *The Cod Fisheries* (Toronto, 1954) for a history of the European and American fisheries of the north Atlantic.
2. Keith Matthews, 'A history of the West of England-Newfoundland fishery' (unpublished DPhil thesis, University of Oxford, 1968) provides a comprehensive analysis of the West Country's role in the trade. See also William B. Stephens 'The West Country ports and the struggle for the Newfoundland fisheries in the 17th century', *Transactions Devonshire Association*, 88 (1956), 90-101, and Neville C. Oswald 'Devon and the cod fishery of Newfoundland', *Transactions Devonshire Association*, 115 (1983), 19-36.
3. See Gillian T. Cell, *English Enterprise in Newfoundland 1577-1660* (Toronto, 1969).
4. Innis, *Cod Fisheries*, 131.
5. Matthews, 'A history of the ... fishery', 29-33.
6. Quoted in D.W. Prowse, *A History of Newfoundland* (New York, 1895), 154-5.
7. This is particularly apparent in Prowse, *History of Newfoundland.*
8. 15 George III, c. 31.

NOTES ON CONTRIBUTORS

Anne DUFFIN, a Cornishwoman, was a post-graduate research student at Exeter University from 1984. She is at present Research Fellow at the Institute of Historical Research, University of London. Her field of study is the political allegiance of the Cornish gentry, c. 1600 to c. 1665.

Robert HIGHAM is Lecturer in medieval archaeology at Exeter University. He has published widely on the archaeology and social history of the medieval castle. He is Secretary of the Centre for South-Western Historical Studies, Secretary of the Castle Studies Group and Chairman of the Devon Archaeological Society.

Valerie MAXFIELD is Senior Lecturer in Roman archaeology at Exeter University, with particular interests in the archaeology of the Roman army and the frontier provinces of the Empire. She has excavated extensively on Roman military sites in England and Scotland and is currently engaged in a field project in Egypt. She is Vice-President, and former President of the Devon Archaeological Society.

Ivan ROOTS is Professor Emeritus at Exeter University, where he was formerly Head of the Department of History and Archaeology. Among his publications, he edited *'Into Another Mould': Aspects of the Interregum* (Exeter Studies in History No. 3, 1980) and *The Monmouth Rising 1685* (1986).

David STARKEY is currently Research Fellow in the Maritime History of Devon in the Departments of History and Archaeology and Economic History at Exeter University. His principal research interests are in maritime elements of British economic and social history. His Exeter PhD thesis (1985) on 'British Privateering, 1702-1783' will shortly be published.

Joyce YOUINGS has recently retired from Exeter University, where she was Professor of English Social History in the Department of History and Archaeology. Her publications include *The Dissolution of the Monasteries* (1971), *Sixteenth-Century England* (1984) and *Raleigh's Century* (1986). She is Chairman of the Councils of the Devonshire Association and of the Devon History Society, and Co-General Editor of the Devon and Cornwall Record Society.